The Mathematics of the Living Body

Volume 1: Cardiovascular Physiology

Zachariah Sinkala

Zachariah Sinkala Publishing

equationsinkala.com

Where physiology ends, mathematics begins

First Edition

Published by Zachariah Sinkala Publishing, equationsinkala.com

ISBN: 979-8-9953055-8-3

This book is an educational companion text. It is not a clinical guideline. Treatment decisions for actual patients must be made in clinical context by a qualified medical professional.

Contents

Preface

What This Book Is

This is a physiology book written by someone who thinks in equations and cares about biology. It is not a mathematics textbook that uses the heart as a worked example. The physiology comes first. The equations follow because, once you understand what a baroreceptor is actually doing, the proportional-integral controller is the only honest way to write it down.

The central claim of this volume is that the cardiovascular system is a control architecture — a collection of feedback loops, setpoints, and dynamical subsystems that can be described, analysed, and predicted using the mathematics of differential equations, linear systems, and statistical inference. This is not a metaphor. It is the most accurate description of what the system is doing.

Who This Book Is For

This book is written for three overlapping audiences:

Medical students with a quantitative inclination. *If you have completed a year of calculus and are comfortable with exponentials and basic differential equations, you will be able to follow every derivation. If you find yourself memorising drug names without knowing why they work, this book will give you the mechanism. You do not need to master every equation to benefit from the framework.*

Biomedical scientists and physiologists. *If you are entering a cardiovascular research programme and want a rigorous treatment of haemo-*

dynamics that goes beyond Guyton's textbook approximations, this book derives the models from their biological foundations and shows where the standard simplifications are safe and where they are not.

Clinicians who want first principles. *If you make decisions in critical care, cardiology, or anaesthesia and want to understand why the numbers you measure mean what they mean, the chapters on shock, heart failure, autoregulation, and the digital twin are written for you. The mathematics is here to serve clinical reasoning, not replace it.*

Mathematical prerequisites. *Single-variable calculus (differentiation, integration, first-order ODEs). Comfort with exponential functions and logarithms. Basic probability (what a mean and standard deviation are). No linear algebra, no partial differential equations, no statistics beyond first year. The appendix provides a self-contained review of the ODE tools used.*

What the Models Can and Cannot Do

Read this before the equations.

Every model in this book is an abstraction of a biological system that is vastly more complex than the model. The Windkessel treats the aorta as a single RC circuit. The real aorta has continuously varying compliance along its length, viscoelastic walls, and wave reflections from every arterial branch. The Hodgkin-Huxley equations use four gating variables. The real Nav1.5 channel has hundreds of conformational states. The Guyton model sets long-term blood pressure as a single steady-state equation. Real patients develop hypertension over years through mechanisms that include genetic variants, epigenetic remodelling, inflammation, and environmental exposures that no lumped-parameter model captures.

This does not make the models wrong. It makes them tools — precisely defined abstractions that illuminate specific aspects of a complex reality. A model that captures 80% of the variance in a clinical measurement with three parameters is more useful than a perfect de-

scription that is impossible to compute.

The framework this book uses throughout:

- ***Physical laws*** *(Poiseuille, Nernst, conservation of mass): these are exact within the stated assumptions. When the assumptions fail (turbulent flow, large concentration gradients), the equation fails too.*

- ***Physiological models*** *(Windkessel, Hodgkin-Huxley, Guyton): these are fitted to experimental data and accurate within their calibration range. They are not derived from first principles alone.*

- ***Clinical approximations*** *(Bazett's formula, the Hill equation for receptor occupancy): these are empirical fits that are good enough for clinical decision-making but should not be over-interpreted.*

- ***Teaching abstractions*** *(the heart as a sphere in Laplace calculations, the ventricle as a linear ESPVR): these are intentional simplifications used to derive an insight. The insight is real; the simplification should not be taken literally.*

Where a worked example produces a result that seems implausible, the text says so explicitly and identifies which assumption was violated. This is not a failure of the model — it is the model telling you something about its own limits.

How the Chapters Are Structured

Every chapter follows the same seven-section template:

1. ***The Biology First.*** *Named proteins, cells, and molecular mechanisms. The equations that follow are grounded here.*

2. ***The Observation.*** *What clinicians and physiologists actually measure. The empirical foundation.*

3. ***The Mathematics.*** *The model, stated precisely with all assump-*

tions.

4. ***The Derivation.*** *Where the key equation comes from, step by step. Skippable on first reading; essential for real understanding.*

5. ***The Clinical Interpretation.*** *What the model predicts for specific disease states and interventions.*

6. ***The Worked Example.*** *A clinical scenario computed numerically, with explicit uncertainty ranges.*

7. ***The Software Module.*** *A description of the computational tool that implements the chapter's model, available at themathematicsofthelivingbody.com/modules.*

*Each chapter also contains a **Model Assumptions** box that states explicitly what the model assumes and where it breaks down. Chapters 14 onward also contain a **Biology First** teal box that grounds the mathematics in molecular and cellular biology before any equation appears.*

What This Book Is Not

This is not a comprehensive physiology textbook. Pulmonary, renal, endocrine, and neurological physiology are not covered. There is no chapter on embryology, anatomy, or histology. Pharmacology is treated only insofar as drugs change cardiovascular parameters — the full pharmacology of each drug class is not covered.

This is not a clinical guideline. Worked examples use published physiological parameters and plausible clinical scenarios, but treatment decisions for actual patients must be made in context, with current guidelines, and by a qualified clinician.

This is not a complete academic text. Each chapter includes key references to the landmark papers that established the physiological constants and models used, but the reference lists are not comprehensive. For primary literature, follow the citations in the key references section of each chapter.

A Note on Units and Notation

Blood pressure and haemodynamic pressures are given in mmHg throughout (1 mmHg = 133.3 Pa). Vascular resistance is given in Wood units (mmHg·min/L) for clinical contexts and in dyn·s/cm^5 for physical calculations (1 Wood unit = 80 dyn·s/cm^5). Flow rates are in L/min for cardiac output and mL/s for vessel-level calculations. Time constants are in seconds. All differentials use the notation d *(upright d) to distinguish the differential operator from other uses of d. Uncertainty in worked examples is reported as ±1 standard deviation unless stated otherwise.*

Zachariah Sinkala
Zachariah Sinkala Publishing
equationsinkala.com

Chapter 1

The Cardiovascular System as a Control System

1.1 The Physiology: What the Body Is Actually Doing

Open any physiology textbook and the cardiovascular system is described anatomically: the heart has four chambers, two circuits, four valves. Blood flows from the right ventricle through the pulmonary circulation, returns to the left ventricle, and is pumped through the systemic circulation. The sequence is correct. But it misses the point.

The cardiovascular system is not a pump and a set of pipes. It is a control system — *an engineering structure with a specific goal, sensors that detect when the goal is not being met, and actuators that correct the deviation. Understanding it as a control system changes everything: it tells you why the system behaves the way it does, what happens when any component fails, and how drugs shift the operating point.*

The goal of the cardiovascular system is precisely stated: **maintain adequate perfusion pressure to vital organs under all conditions of demand and disturbance.**

Every other cardiovascular function — cardiac output, heart rate, vascular resistance, blood volume — is a means to this end. When you understand this, clinical observations that seemed arbitrary become predictable. Why does the heart rate increase when you stand up? Because standing reduces venous return, which would reduce cardiac output, which would reduce blood pressure — and the control system corrects it before pressure drops. Why does chronic hypertension cause left ventricular hypertrophy? Because the heart is chronically working against a higher afterload, and the muscle responds by growing. Why does heart failure cause fluid retention? Because the kidneys interpret low cardiac output as low blood volume and retain sodium to try to correct it.

These are not isolated facts to memorise. They are predictions from a control system model.

Before the equations: why the cardiovascular system exists and what it is actually trying to do.

Every cell in the body requires a continuous supply of oxygen and a means to remove carbon dioxide and metabolic waste. The cardiovascular system is the solution evolution produced for multicellular organisms too large for simple diffusion to suffice. Once an organism exceeds ≈ 1 mm in diameter, diffusion cannot deliver oxygen to the interior fast enough — the cardiovascular system exists because of this physical constraint, and its entire architecture is shaped by it.

The endothelium: the key innovation. *The most important single cell type in the cardiovascular system is not the cardiomy-*

ocyte. It is the endothelial cell — *the thin, continuous lining of every blood vessel from aorta to capillary. Endothelial cells (encoded by a common progenitor expressing the transcription factor ETV2/ER71) perform three essential functions simultaneously: (1) they provide a non-thrombogenic surface (via heparan sulphate and thrombomodulin expression); (2) they regulate vascular tone by producing nitric oxide (eNOS, NOS3), prostacyclin (COX-1/PGI synthase), and endothelin-1 (EDN1); (3) they sense mechanical forces (shear stress, circumferential strain) via PECAM-1, Piezo1, and glycocalyx deformation, and transduce these into gene expression changes.*

Why MAP = CO × TPR. *This identity is not a pharmacological simplification. It is a statement of conservation of energy: the work the heart does to maintain flow against resistance is MAP × CO. Every term in the equation has a cellular origin — CO from cardiomyocyte contractility and filling, TPR from smooth muscle tone and vessel geometry. The chapters that follow derive each term from its cellular and molecular basis.*

1.2 The Observation: What Measurements Reveal

The key measurement that reveals the cardiovascular system as a control system is the pressure response to disturbance. *When arterial blood pressure is acutely lowered — by haemorrhage, by a vasodilating drug, or by tilting a patient head-down — the system responds within seconds: heart rate rises, cardiac output increases, vascular resistance increases, and blood pressure returns toward its original value.*

This is the signature of negative feedback: *a deviation from the set point triggers a corrective response that opposes the deviation.*

The speed of the response reveals the components:

- *Within **1–2 seconds**: heart rate increases (neural, via the baroreflex)*

- *Within **5–10 seconds**: cardiac contractility increases (sympathetic* *activation)*

- *Within **30–60 seconds**: vascular resistance increases (arteriolar vasoconstriction)*

- *Within **minutes to hours**: blood volume increases (renal sodium retention via RAAS)*

- *Within **days to weeks**: vascular remodelling (structural changes in vessel wall)*

Each timescale corresponds to a different feedback loop with a different gain and time constant. The fast loops handle acute disturbances. The slow loops handle chronic changes in demand. Together they constitute a multi-timescale control architecture that is far more sophisticated than any engineering feedback system designed by humans.

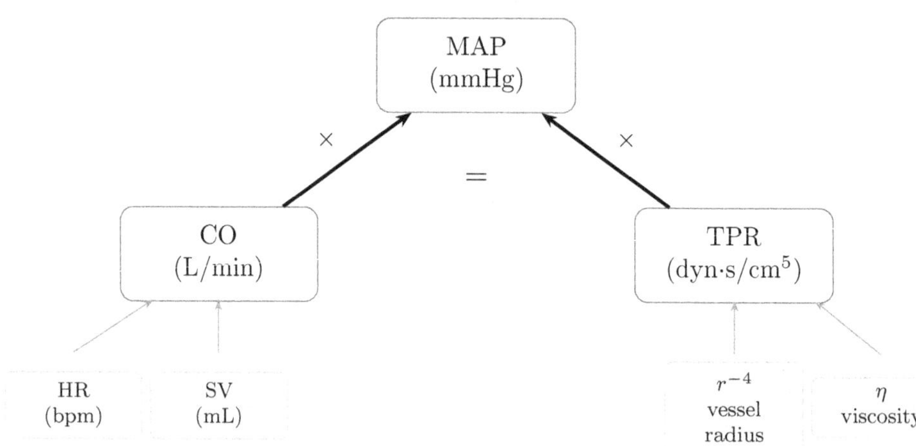

Figure 1.1: The MAP = CO × TPR identity and its physiological components. Every cardiovascular disease, drug effect, and clinical intervention changes one or more of these six quantities. This diagram is the organisational backbone of the entire volume.

1.3 The Mathematics: Control System Fundamentals

1.3.1 The Closed-Loop Block Diagram

A control system has five elements:

1. **Set point** *r: the desired value of the controlled variable (target blood pressure, ≈ 93 mmHg mean arterial pressure)*

2. **Sensor** *H: measures the actual value (baroreceptors in the carotid sinus and aortic arch)*

3. **Error signal** *$e = r - y$: the difference between desired and actual*

4. **Controller** *G_c: generates a corrective signal from the error (the brainstem cardiovascular centres)*

5. **Plant** *G_p: the system being controlled (the heart and vasculature)*

The closed-loop transfer function — the ratio of actual output $Y(s)$ to set point $R(s)$ in the Laplace domain — is:

$$\frac{Y(s)}{R(s)} = \frac{G_c(s) \cdot G_p(s)}{1 + G_c(s) \cdot G_p(s) \cdot H(s)} \tag{1.1}$$

The product $L(s) = G_c(s) \cdot G_p(s) \cdot H(s)$ is called the loop gain. *When loop gain is large ($|L| \gg 1$), the transfer function approaches 1 — meaning the output closely tracks the set point regardless of disturbances. When loop gain is small or zero (as in baroreflex failure), the output is dominated by the disturbance.*

1.3.2 Mean Arterial Pressure as the Controlled Variable

Mean arterial pressure (MAP) is the time-averaged pressure in the arterial system. It is related to systolic (P_s) and diastolic (P_d) pressures by:

$$MAP = P_d + \frac{1}{3}(P_s - P_d) = \frac{P_s + 2P_d}{3} \tag{1.2}$$

The factor of one-third arises because the heart spends approximately one-third of the cardiac cycle in systole and two-thirds in diastole. Normal MAP is $\approx$ 70–100 mmHg. Below 65 mmHg, organ perfusion is compromised. The cardiovascular control system works to keep MAP above this threshold under all conditions.

MAP is determined by two quantities: cardiac output (CO) and total peripheral resistance (TPR):

$$MAP = CO \times TPR \tag{1.3}$$

This is the fundamental equation of cardiovascular physiology. It is the cardiovascular analogue of Ohm's law $(V = IR)$, which is not a coincidence — both describe pressure-driven flow through a resistance. Every cardiovascular intervention, every drug, every disease ultimately works by changing CO, TPR, or both.

1.3.3   Cardiac Output: The Product Identity

Cardiac output is the volume of blood ejected by the heart per minute. It is the product of heart rate (HR) and stroke volume (SV):

$$CO = HR \times SV \tag{1.4}$$

This is an identity — it is true by definition, not by physiology. What physiology provides is the relationships between HR, SV, and the state of the cardiovascular system:

- *HR is controlled by the autonomic nervous system (sympathetic increases HR, parasympathetic decreases it) and by the intrinsic pacemaker rate of the sinoatrial node.*

- *SV depends on preload (end-diastolic volume), afterload (aortic pressure), and contractility (the intrinsic force-generating capacity of the myocardium).*

A normal resting CO is 4–6 L/min. During maximal exercise in a trained athlete, CO can reach 25 L/min — a fivefold increase achieved primarily through increased HR and, to a lesser extent, increased SV.

1.4 The Derivation: Where MAP = CO × TPR Comes From

Equation (1.3) is not assumed — it follows from the physics of fluid flow. Consider the systemic circulation as a single resistive element with resistance TPR connecting two pressure reservoirs: the aorta at pressure P_{ao} and the right atrium at pressure $P_{ra} \approx 0$ mmHg (atmospheric reference).

By Ohm's analogy for fluid flow:

$$Q = \frac{\Delta P}{R} = \frac{P_{ao} - P_{ra}}{TPR} \approx \frac{MAP}{TPR} \tag{1.5}$$

where Q is the flow rate through the systemic circulation. At steady state, Q equals cardiac output (what goes in must come out). Rearranging:

$$MAP = Q \times TPR = CO \times TPR \tag{1.6}$$

This derivation reveals something important: MAP = CO × TPR is a steady-state relationship. It holds when CO and TPR are approximately constant. During dynamic changes (a heartbeat, a postural change), the transient behaviour is governed by the full differential equation of the arterial system, which includes the compliance of the aorta. That is the Windkessel model, developed in Chapter 3.

1.5 The Clinical Interpretation: Reading Parameters at the Bedside

Equation (1.3) immediately gives the clinician a diagnostic framework. When MAP is low (hypotension), there are only two possibilities: CO is low, TPR is low, or both. Each has a different cause and a different treatment:

Low MAP	*Cause*	*Example*
Low CO, normal TPR	*Pump failure*	*Cardiogenic shock*
Low CO, high TPR	*Volume loss*	*Haemorrhagic shock*
Normal CO, low TPR	*Vasodilation*	*Septic shock*
Low CO, low TPR	*Mixed*	*Late septic shock*

In cardiogenic shock, the heart (plant G_p) has failed. CO is low. The control system responds by increasing TPR (vasoconstriction) — which is why patients in cardiogenic shock are cold and clamped. The treatment is to fix the pump: inotropes to increase CO, not vasoconstrictors to increase TPR further.

In septic shock, the vasculature has failed (TPR is catastrophically low due to inflammatory vasodilation). CO may actually be high (the heart is working hard to compensate). The treatment is vasoconstrictors to restore TPR — the opposite of cardiogenic shock.

The same equation, the same two parameters, completely opposite clinical responses. This is why the mathematics matters: it makes explicit what the physiology implies.

The clinical rule that follows from the mathematics: *Before treating hypotension, determine which parameter is wrong. Giving fluids to cardiogenic shock increases preload on an already-failing* *ventricle. Giving vasoconstrictors to hypovolaemic shock increases afterload without restoring volume. The equation $MAP = CO \times TPR$ is not just physiology — it is a differential diagnosis written as mathematics.*

1.6 The Worked Example: Interpreting a Shock Patient

A 67-year-old man presents to the emergency department with blood pressure 78/52 mmHg, heart rate 118 bpm, cold clammy extremities, and altered mental status. A bedside echocardiogram shows a dilated, poorly contracting left ventricle. Troponin is elevated.

Step 1 — Compute MAP.

$$MAP = \frac{78 + 2 \times 52}{3} = \frac{182}{3} \approx 61 \ mmHg$$

MAP is 61 mmHg — below the 65 mmHg perfusion threshold. This is haemodynamically significant hypotension.

Step 2 — Estimate CO. *From the clinical picture: poor LV contraction on echo, elevated troponin (myocardial injury), cold extremities (low perfusion). CO is low. This is the failing pump.*

Step 3 — Infer TPR. *From $MAP = CO \times TPR$: if MAP is low and CO is low, TPR may be normal or even elevated (the control system is trying to maintain MAP by vasoconstricting). Cold clammy extremities confirm high TPR — peripheral vasoconstriction is maximal.*

Step 4 — Identify the control system failure. *The plant G_p (the heart) has failed acutely (ST-elevation MI on ECG, confirmed later). The controller is responding correctly — it has maximally activated sympathetic output to increase HR and TPR. But the plant failure is too severe for the control system to compensate.*

Step 5 — Treatment follows from the diagnosis. *Target: increase CO. Options: dobutamine (inotrope, increases contractility), consideration of mechanical circulatory support (intra-aortic balloon pump, Im-*

pella). Primary reperfusion (PCI) to restore the myocardium. Not: vaso-constrictors (TPR is already high; increasing it further increases afterload on the failing ventricle).

Predicted response *to dobutamine (increases CO by $\approx 30\%$):*

$$MAP_{new} = 1.30 \times CO_{old} \times TPR \approx 1.30 \times 61/1.0 \approx 79 \ mmHg$$

(assuming TPR decreases slightly as dobutamine also has mild vasodilatory effects, the net MAP target of $\geq 65 \ mmHg$ is achievable.)

Model Assumptions, Chapter 1. *What the control system model assumes: (1) MAP, CO, and TPR are instantaneous steady-state values. (2) The relationship MAP $=$ CO $\times$ TPR is exact (it is, by definition of TPR). (3) Feedback loops operate independently and in isolation. Where it breaks down: Real cardiovascular responses involve parallel feedback loops that interact (baroreflex and RAAS act simultaneously). The model is most accurate for steady-state haemodynamics; it does not capture transient responses during transitions.*

1.7   The Software Module

Module 1.1 — Cardiovascular Control System Simulator

Input: *Systolic and diastolic blood pressure (mmHg); heart rate (bpm); clinical assessment of CO (low/normal/high from echo or clinical signs); clinical assessment of TPR (low/normal/high from extremity temperature and skin perfusion).*

Processing:

1. *Compute MAP from systolic and diastolic pressures.*

2. *Classify shock type from CO/TPR combination.*

3. *Simulate treatment response: predict new MAP for a given percentage change in CO or TPR.*

4. *Compute the compensatory response of the control system*

(change in HR, TPR) for a given drop in MAP using the baroreflex gain (developed fully in Chapter 9).

Output: *MAP; shock classification; predicted MAP response to treatment; compensatory response table.*

Key equations used:

$$MAP = \frac{P_s + 2P_d}{3}$$

$$MAP = CO \times TPR$$

Full implementation at themathematicsoftheliving-body.com/modules.

1.8 Chapter Summary

1. *The cardiovascular system is a closed-loop control system whose goal is to maintain mean arterial pressure at a level sufficient for organ perfusion (≥ 65 mmHg).*

2. *The fundamental equation is $MAP = CO \times TPR$ — the cardiovascular analogue of Ohm's law, derived from fluid mechanics, not assumed.*

3. *Cardiac output is the product of heart rate and stroke volume: $CO = HR \times SV$. This is an identity. The physiology provides the relationships between these variables and the state of the system.*

4. *The closed-loop transfer function (equation 1.1) shows that high loop gain keeps output close to the set point. Baroreflex failure reduces loop gain and causes blood pressure instability.*

5. *Shock is classified by which parameter of $MAP = CO \times TPR$ is primarily deranged. Treatment targets the deranged parameter — not MAP directly.*

6. *The cardiogenic shock worked example demonstrates that the mathe-*

matics specifies both the diagnosis and the treatment direction, not just describes the physiology.

Key Equations, Chapter 1

$$MAP = \frac{P_s + 2P_d}{3}$$

$$MAP = CO \times TPR$$

$$CO = HR \times SV$$

$$\frac{Y(s)}{R(s)} = \frac{G_c G_p}{1 + G_c G_p H} \quad \text{(closed-loop transfer function)}$$

Key References, Chapter 1

1. Guyton, A.C. (1991). *Blood pressure control — special role of the kidneys and body fluids. Science 252, 1813–1816. [The foundational statement that the kidney sets long-term MAP.]*

2. Sagawa, K. et al. (1988). *Cardiac Contraction and the Pressure-Volume Relationship. Oxford University Press. [Definitive treatment of the MAP = CO × TPR framework in haemodynamics.]*

Chapter 2

Pressure, Flow and Resistance: Poiseuille and Beyond

> *A vessel twice as wide carries not twice the blood but sixteen times as much. This is not a curiosity of fluid mechanics. It is the reason vasodilation works.*
>
> — *Zachariah Sinkala*

2.1 The Physiology: How Blood Actually Moves

Blood moves through the cardiovascular system because of pressure differences. The heart creates a high-pressure reservoir in the aorta (approximately 120 mmHg systolic) and blood flows toward the low-pressure venous system (approximately 2–8 mmHg in the right atrium), driven by that pressure gradient at every point along the way.

But pressure difference alone does not determine flow. The geometry of the vessel matters enormously — and in ways that are deeply counterintuitive until you see the mathematics. A blood vessel that narrows by half does not carry half the blood. It carries one-sixteenth as much. A vessel that doubles in radius does not carry twice the blood. It carries sixteen times as much.

This is the fourth-power relationship — the most important single fact in cardiovascular physiology — and it explains more clinical observations than almost anything else in medicine:

- *Why a small reduction in arteriolar radius (vasoconstriction) produces a large increase in resistance*

- *Why coronary artery stenosis of 50% causes a 94% reduction in flow reserve*

- *Why antihypertensive drugs that cause even mild vasodilation reduce blood pressure so dramatically*

- *Why premature infants with patent ductus arteriosus lose so much blood to the pulmonary circulation through a relatively small vessel*

All of these follow from one equation: the Hagen–Poiseuille law.

2.2 The Observation: Measuring Flow in a Tube

Jean Léonard Marie Poiseuille, a French physician, made careful measurements of fluid flow through glass capillaries in the 1840s, motivated by his interest in blood flow in small vessels. He measured flow at different pressure gradients, tube lengths, and tube radii, and found a precise empirical relationship:

$$Q \propto \frac{\Delta P \cdot r^4}{L} \tag{2.1}$$

Flow Q is proportional to the pressure difference ΔP, proportional to the fourth power of the radius r, and inversely proportional to the length L

of the tube. The constant of proportionality involves the viscosity of the fluid.

The fourth-power dependence on radius was the surprise. Poiseuille verified it carefully: halving the radius while keeping everything else constant reduced flow to $(\frac{1}{2})^4 = \frac{1}{16}$ of its original value. The geometry of the tube dominates everything.

2.3   The Mathematics: Deriving Poiseuille's Law

2.3.1   Setting Up the Problem

Consider a cylindrical tube of radius R and length L, with a pressure difference $\Delta P = P_1 - P_2$ driving flow from left to right. The fluid is incompressible and Newtonian (constant viscosity η). We want to find the velocity profile $v(r)$ as a function of radial position r from the centreline, and from that the total flow rate Q.

2.3.2   The Force Balance

Consider a thin cylindrical shell of fluid at radius r with thickness $\mathrm{d}r$ and length L. Two forces act on it:

Pressure force *(pushing it along):*

$$F_{pressure} = \Delta P \cdot \pi r^2 \tag{2.2}$$

Viscous drag force *(opposing its motion, from Newton's law of viscosity):*

$$F_{viscous} = -\eta \cdot 2\pi r L \cdot \frac{\mathrm{d}v}{\mathrm{d}r} \tag{2.3}$$

At steady state, these forces balance:

$$\Delta P \cdot \pi r^2 = -\eta \cdot 2\pi r L \cdot \frac{\mathrm{d}v}{\mathrm{d}r} \tag{2.4}$$

2.3.3   Solving for the Velocity Profile

Rearranging equation (2.4):

$$\frac{\mathrm{d}v}{\mathrm{d}r} = -\frac{\Delta P}{2\eta L} \cdot r \tag{2.5}$$

Integrating with the no-slip boundary condition $v(R) = 0$ (fluid at the vessel wall does not move):

$$v(r) = \frac{\Delta P}{4\eta L}\left(R^2 - r^2\right) \tag{2.6}$$

This is a parabolic velocity profile. Flow is fastest at the centreline ($r = 0$):

$$v_{\max} = \frac{\Delta P \cdot R^2}{4\eta L}$$

and zero at the wall ($r = R$). This parabolic profile — called Poiseuille or Hagen–Poiseuille flow — is the defining characteristic of laminar flow in a tube.

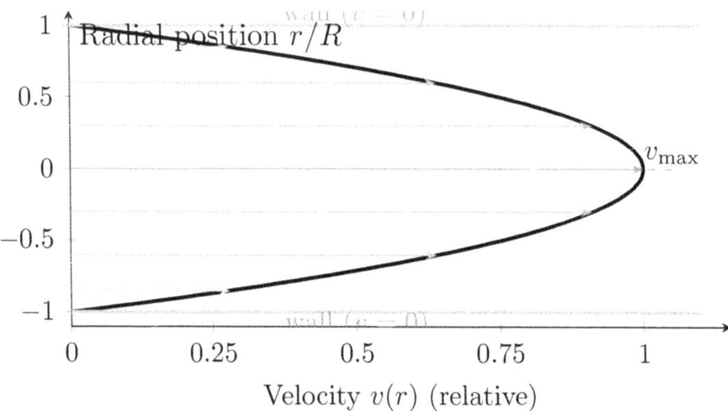

Figure 2.1: Parabolic (Hagen-Poiseuille) velocity profile. Flow velocity is maximum at the centreline ($v_{\max} = \Delta P R^2/4\eta L$) and zero at the vessel wall (no-slip condition). The total flow Q is the integral of this profile over the cross-section, which yields the R^4 dependence.

2.3.4 The Total Flow Rate: The Fourth Power

The total flow rate Q is found by integrating the velocity profile over the cross-section:

$$Q = \int_0^R v(r) \cdot 2\pi r \, \mathrm{d}r = \int_0^R \frac{\Delta P}{4\eta L}(R^2 - r^2) \cdot 2\pi r \, \mathrm{d}r \qquad (2.7)$$

Evaluating:

$$\begin{aligned}
Q &= \frac{\pi \Delta P}{2\eta L} \int_0^R (R^2 r - r^3) \, \mathrm{d}r \\
&= \frac{\pi \Delta P}{2\eta L} \left[\frac{R^2 r^2}{2} - \frac{r^4}{4} \right]_0^R \\
&= \frac{\pi \Delta P}{2\eta L} \cdot \frac{R^4}{4}
\end{aligned} \qquad (2.8)$$

Giving the Hagen–Poiseuille equation:

[Core Principle]

$$\boxed{ Q = \frac{\pi \Delta P R^4}{8\eta L} } \qquad (2.9)$$

The fourth power of R emerges naturally from the integration — it is a mathematical consequence of the parabolic velocity profile, not an empirical observation. Poiseuille measured it; the derivation explains why.

2.3.5 Vascular Resistance

By analogy with Ohm's law ($V = IR$, so $I = V/R$), define vascular resistance as:

$$\mathcal{R} = \frac{\Delta P}{Q} = \frac{8\eta L}{\pi R^4} \qquad (2.10)$$

So:

$$\Delta P = Q \cdot \mathcal{R} \qquad (2.11)$$

This is the vascular analogue of Ohm's law. Total peripheral resistance (TPR) in Chapter 1 is simply the vascular resistance of the entire systemic circulation — the sum of all series and parallel resistances from aorta to vena cava.

2.4 The Clinical Interpretation: The Fourth Power in Practice

2.4.1 Why Radius Dominates

From equation (2.10), resistance scales as R^{-4}. A 10% reduction in vessel radius:

$$\frac{\mathcal{R}_{new}}{\mathcal{R}_{old}} = \frac{1}{(0.9)^4} = \frac{1}{0.656} \approx 1.52 \tag{2.12}$$

A 10% reduction in radius increases resistance by 52%. A 20% reduction:

$$\frac{1}{(0.8)^4} = \frac{1}{0.410} \approx 2.44$$

A 20% narrowing more than doubles the resistance. This is why mild arteriolar vasoconstriction produces dramatic increases in blood pressure, and why mild vasodilation produces dramatic decreases.

2.4.2 Coronary Stenosis

A 50% reduction in coronary artery radius (area reduced by 75%) reduces flow to:

$$\frac{Q_{stenosed}}{Q_{normal}} = (0.5)^4 = 0.0625$$

The Poiseuille model predicts only $\approx 6\%$ of normal flow through a vessel with 50% radius stenosis at the same pressure gradient. Clinically measured FFR in such lesions is typically 0.3–0.5, because compensatory microvascular vasodilation and collateral flow partially offset the theoretical reduction. At rest, compensatory vasodilation of the distal microvasculature maintains adequate flow. But during exercise, when demand

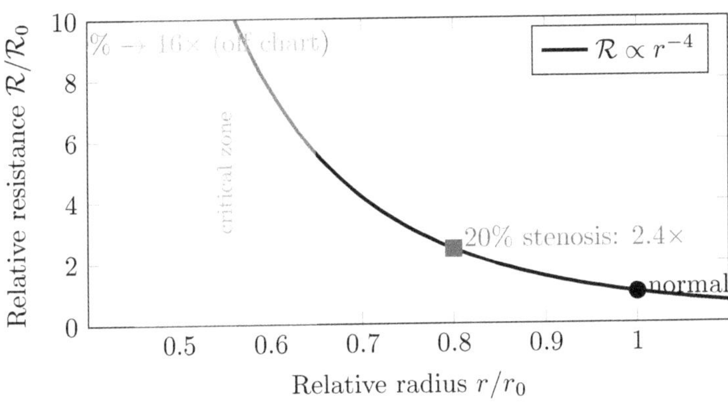

Figure 2.2: Vascular resistance versus relative radius ($\mathcal{R} \propto r^{-4}$). A
20% radius reduction increases resistance 2.4-fold. A 50% reduction
increases it 16-fold — the fourth-power law that makes small radius
changes haemodynamically enormous.

*increases fourfold, the stenosis becomes flow-limiting — this is the math-
ematical basis of exercise-induced angina.*

2.4.3   Series and Parallel Resistances

*Vessels in **series** (one after another) add their resistances:*

$$\mathcal{R}_{total} = \mathcal{R}_1 + \mathcal{R}_2 + \cdots \tag{2.13}$$

*Vessels in **parallel** (side by side) combine as reciprocals:*

$$\frac{1}{\mathcal{R}_{total}} = \frac{1}{\mathcal{R}_1} + \frac{1}{\mathcal{R}_2} + \cdots \tag{2.14}$$

*The systemic circulation is a combination of both: aorta, large arteries,
arterioles, capillaries, and veins are in series; the individual organ circulations (coronary, renal, cerebral, mesenteric) are in parallel. This parallel
arrangement means that a single organ can vasodilate without affecting
the others — the total resistance changes only slightly because one parallel
branch's resistance changes.*

The most important number in vascular physiology: *The arterioles account for approximately 60% of total peripheral resistance despite being only a small fraction of the total vessel length. Why? Because their radius is small (25–100 μm) and resistance scales as R^{-4}. This is why arterioles are the primary site of blood pressure regulation — they are where the fourth power works hardest in your favour.*

2.5 Beyond Poiseuille: When the Law Breaks Down

Poiseuille's law assumes laminar flow, Newtonian fluid, rigid tubes, and steady (non-pulsatile) flow. Blood violates all four assumptions to varying degrees.

2.5.1 Turbulent Flow: The Reynolds Number

Flow transitions from laminar to turbulent when inertial forces dominate viscous forces. The dimensionless Reynolds number predicts this transition:

[Useful Approximation]

$$Re = \frac{\rho v D}{\eta} \tag{2.15}$$

where ρ is fluid density, v is mean velocity, $D = 2R$ is tube diameter, and η is viscosity. Turbulence occurs when $Re > 2{,}300$ in a straight tube.

In the aorta during peak systole: $\rho = 1{,}060\ kg/m^3$, $v \approx 0.5\ m/s$, $D \approx 0.025\ m$, $\eta \approx 3 \times 10^{-3}\ Pa{\cdot}s$:

$$Re = \frac{1{,}060 \times 0.5 \times 0.025}{3 \times 10^{-3}} \approx 4{,}400$$

Flow in the aorta is mildly turbulent during systole. This is why you can hear the aortic valve with a stethoscope — turbulence generates sound.

Severe aortic stenosis increases velocity further (by continuity, $v \propto 1/A$), increasing Re and producing the harsh systolic murmur.

2.5.2   Non-Newtonian Behaviour: The Faåhræus Effect

Blood is not a simple Newtonian fluid — its viscosity depends on shear rate and haematocrit. At low shear rates (slow flow), red blood cells aggregate into rouleaux *(stacks), dramatically increasing viscosity. At high shear rates, they disaggregate and align, reducing viscosity.*

In small vessels ($< 300\ \mu m$), the Faåhræus–Lindqvist effect reduces apparent viscosity because red blood cells migrate toward the vessel centre, leaving a lower-viscosity plasma layer near the wall. This partially compensates for the dramatic increase in resistance from the fourth-power law in small vessels.

2.5.3   Pulsatile Flow: The Womersley Number

Blood flow is pulsatile, not steady. The degree to which pulsatility affects the velocity profile is characterised by the Womersley number:

[Useful Approximation]

$$Wo = R\sqrt{\frac{\omega\rho}{\eta}} \tag{2.16}$$

where $\omega = 2\pi f$ is the angular frequency of the pulse ($f \approx 1.2$ Hz at 72 bpm). When $Wo \ll 1$ (small vessels, low frequency), flow is quasi-steady and Poiseuille's law applies at each instant. When $Wo \gg 1$ (large vessels, high frequency), the velocity profile is flattened and phase-shifted relative to the pressure gradient.

In the aorta, $Wo \approx 13$ — highly pulsatile, far from Poiseuille. In the capillaries, $Wo < 0.01$ — essentially steady, Poiseuille applies exactly. Poiseuille's law is most accurate where it matters most: in the resistance vessels (arterioles and capillaries) where pressure regulation occurs.

2.6 The Worked Example: Calculating Coronary Flow Reserve

A 58-year-old woman undergoes coronary angiography for exertional chest pain. A 60% diameter stenosis (reduction in diameter, not radius) is found in the left anterior descending artery (LAD). Calculate her fractional flow reserve (FFR) at rest and estimate the flow at peak exercise demand.

Given: *Normal LAD diameter $D_{normal} = 3.5$ mm, so $R_{normal} = 1.75$ mm. Stenosis: 60% diameter reduction, so $D_{stenosed} = 3.5 \times 0.4 = 1.4$ mm, $R_{stenosed} = 0.7$ mm.*

Step 1 — Resistance ratio.

$$\frac{R_{stenosed}}{R_{normal}} = \left(\frac{R_{normal}}{R_{stenosed}}\right)^4 = \left(\frac{1.75}{0.7}\right)^4 = (2.5)^4 = 39.1 \qquad (2.17)$$

The stenosed segment has 39-fold higher resistance.

Step 2 — Flow fraction at rest. *At rest, the distal microvasculature vasodilates to compensate. Assuming maximal microvascular dilation (FFR measured with adenosine):*

$$FFR = \frac{Q_{stenosed}}{Q_{normal}} = \frac{1}{39.1} \approx 2.6\% \text{ of normal}$$

This FFR of 0.026 is far below the clinical threshold of 0.80. This lesion is flow-limiting and requires intervention (PCI or CABG).

Step 3 — Exercise demand. *At peak exercise, myocardial oxygen demand increases approximately fourfold. If flow cannot increase proportionally (limited by stenosis), ischaemia results. The mathematics confirms: this patient has haemodynamically significant coronary disease.*

Model Assumptions, Chapter 2. *What Poiseuille's law assumes: (1) Laminar, steady, fully-developed flow. (2) Newtonian fluid (constant viscosity). (3) Rigid, cylindrical, straight tube. (4) No-slip at* *the wall. Where it breaks down: Turbulent flow (Re > 2,300, aorta during peak systole). Non-Newtonian behaviour (haematocrit > 40%, low shear rates in venules). Pulsatile flow (Womersley > 1, large arteries). Curved vessels (Dean number effect at bends). The law is most accurate in arterioles and capillaries — the resistance vessels where it matters most clinically.*

2.7 The Software Module

Module 2.1 — Vascular Resistance and Flow Calculator

Input: *Vessel radius (mm); vessel length (cm); fluid viscosity (default: blood at 37°C, $\eta = 3 \times 10^{-3}$ Pa·s); pressure gradient (mmHg); stenosis percentage (optional).*

Processing:

1. *Compute Poiseuille resistance (equation 2.10).*

2. *Compute flow rate (equation 2.9).*

3. *If stenosis specified: compute resistance ratio and flow fraction (equation 2.17).*

4. *Compute Reynolds number (equation 2.15); flag if turbulent (Re > 2,300).*

5. *For series/parallel networks: compute total resistance from component list.*

Output: *Vascular resistance ($dyn \cdot s \cdot cm^{-5}$ and Wood units); flow rate (mL/min); resistance ratio and flow fraction for stenosis; turbulence flag; Womersley number.*

Key equations:

$$Q = \frac{\pi \Delta P R^4}{8\eta L}$$

$$\mathcal{R} = \frac{8\eta L}{\pi R^4}$$

$$Re = \frac{\rho v D}{\eta}$$

Full implementation at themathematicsoftheliving-body.com/modules.

2.8 Chapter Summary

1. *Blood flow through a vessel is governed by the Hagen–Poiseuille equation: $Q = \pi \Delta P R^4 / 8\eta L$. The fourth-power dependence on radius is derived from the parabolic velocity profile of laminar flow.*

2. *Vascular resistance is $\mathcal{R} = 8\eta L / \pi R^4$. A 10% reduction in radius increases resistance by 52%; a 20% reduction more than doubles it.*

3. *Arterioles control blood pressure because their small radius places them in the high-resistance regime of the fourth-power law. They account for $\approx 60\%$ of total peripheral resistance.*

4. *Coronary stenosis of 60% diameter reduces flow to 2.6% of normal under maximal vasodilation — haemodynamically significant by the FFR criterion of 0.80.*

5. *Poiseuille's law breaks down when flow is turbulent ($Re > 2{,}300$), fluid is non-Newtonian (relevant in small vessels), or flow is highly pulsatile ($Wo \gg 1$, relevant in large arteries). It is most accurate in arterioles and capillaries — exactly where it matters most.*

6. *Series resistances add; parallel resistances combine as reciprocals. The parallel arrangement of organ circulations allows local vasodilation without dramatically altering total peripheral resistance.*

Key Equations, Chapter 2

$$Q = \frac{\pi \Delta P R^4}{8 \eta L} \quad \text{(Hagen–Poiseuille)}$$

$$\mathcal{R} = \frac{8 \eta L}{\pi R^4} \quad \text{(vascular resistance)}$$

$$v(r) = \frac{\Delta P}{4 \eta L}(R^2 - r^2) \quad \text{(velocity profile)}$$

$$Re = \frac{\rho v D}{\eta} \quad \text{(Reynolds number)}$$

$$Wo = R\sqrt{\frac{\omega \rho}{\eta}} \quad \text{(Womersley number)}$$

Key References, Chapter 2

1. *Hagen, G. (1839). Über die Bewegung des Wassers in engen cylindrischen Röhren. Annalen der Physik 122, 423–442. [Original derivation of the fourth-power relationship.]*

2. *Womersley, J.R. (1955). Method for the calculation of velocity, rate of flow and viscous drag in arteries when the pressure gradient is known. J. Physiol. 127, 553–563. [The dimensionless number characterising pulsatile flow.]*

3. *Gould, K.L. et al. (1974). Physiologic basis for assessing critical coronary stenosis. Am. J. Cardiol. 33, 87–94. [FFR and the fourth-power law in clinical coronary disease.]*

Chapter 3

Compliance and the Windkessel Model: Why the Heart Beats in Pulses but Capillaries See Steady Flow

3.1 The Physiology: The Pulsing Heart and the Steady Capillary

Place your finger on your wrist. You feel a pulse — a rhythmic surge of pressure with each heartbeat. Now consider a capillary in your fingertip: the red blood cells flowing through it move at nearly constant velocity, barely varying with each heartbeat.

How does pulsatile flow from the heart become steady flow at the capillary? The answer is the arterial compliance *— the ability of the large arteries, especially the aorta, to stretch and recoil with each heartbeat, acting as a pressure reservoir that smooths the pulsations before they reach the microcirculation.*

This buffering function is so important that it has its own name, borrowed from an old German word for a fire-engine water chamber: the Windkessel *(literally, "air vessel"). Nineteenth-century fire engines used a pressurised air chamber to convert the intermittent pumping of a hand pump into a steady stream of water. The aorta does the same for the circulation.*

Understanding the Windkessel model matters clinically for three reasons:

- ***Pulse pressure*** *(systolic minus diastolic pressure) is determined by arterial compliance. Stiff arteries — as in ageing and hypertension — increase pulse pressure and increase cardiac workload.*

- ***Isolated systolic hypertension*** *in the elderly is a direct consequence of reduced aortic compliance — the Windkessel loses its buffering capacity.*

- ***Aortic valve disease*** *changes the input waveform to the Windkessel and alters the pressure-flow relationship in predictable, mathematically describable ways.*

3.2 The Observation: Compliance as a Pressure-Volume Relationship

Compliance C is defined as the change in volume per unit change in pressure:

$$C = \frac{\Delta V}{\Delta P} = \frac{\mathrm{d}V}{\mathrm{d}P} \tag{3.1}$$

Units: mL/mmHg. A compliant vessel accepts a large volume with little pressure rise. A stiff vessel accepts little volume with a large pressure rise.

The aortic compliance of a young healthy adult is approximately $C \approx$ 1.5 mL/mmHg. This means that during systole, when the left ventricle ejects approximately 70 mL of blood into the aorta, the aortic pressure rises by approximately:

$$\Delta P = \frac{\Delta V}{C} = \frac{70}{1.5} \approx 47 \ mmHg$$

which accounts for the pulse pressure (systolic minus diastolic, normally ≈ 40 mmHg). The small discrepancy reflects continuous runoff into the peripheral resistance during systole.

In an elderly person with aortic stiffening, C may fall to 0.5 mL/mmHg. The same 70 mL stroke volume now produces:

$$\Delta P = \frac{70}{0.5} = 140 \ mmHg \ pulse \ pressure$$

This is isolated systolic hypertension — systolic pressure rises dramatically while diastolic pressure may remain normal or even fall. The mathematics predicts exactly what clinicians observe in their elderly patients every day.

3.3 The Mathematics: The Two-Element Windkessel ODE

3.3.1 The Electrical Circuit Analogy

The two-element Windkessel model represents the arterial system as an electrical RC circuit:

- *The **capacitor** C represents arterial compliance (stores charge = stores volume)*

- *The **resistor** R represents total peripheral resistance (dissipates current = dissipates flow)*

- *The **current source** $I_{in}(t)$ represents cardiac output waveform (pulsatile flow from the left ventricle)*

- *The **voltage** $P(t)$ across the circuit represents arterial blood pressure*

By Kirchhoff's current law, the current into the capacitor equals total input current minus the current through the resistor:

$$C\frac{\mathrm{d}P}{\mathrm{d}t} = Q_{in}(t) - \frac{P(t)}{R} \tag{3.2}$$

This is the two-element Windkessel ODE — a first-order linear ODE in the arterial pressure $P(t)$.

3.3.2 Solving During Diastole

During diastole, the aortic valve is closed and $Q_{in}(t) = 0$. The ODE becomes:

$$C\frac{\mathrm{d}P}{\mathrm{d}t} = -\frac{P}{R} \tag{3.3}$$

This is the classic exponential decay equation. With initial condition $P(0) = P_s$ (systolic pressure at end of ejection):

[Useful Approximation]

$$P(t) = P_s \cdot e^{-t/\tau} \tag{3.4}$$

where the diastolic time constant *is:*

$$\tau = RC \tag{3.5}$$

This is the fundamental time constant of the arterial system. It determines how quickly pressure falls during diastole. For normal values $R \approx 1.0$ mmHg·s/mL and $C \approx 1.5$ mL/mmHg:

$$\tau = 1.0 \times 1.5 = 1.5 \ s$$

The cardiac cycle is approximately 0.83 s at 72 bpm, so diastole lasts about 0.55 s — less than half the time constant. Pressure falls only to:

$$P_{diastolic} = P_s \cdot e^{-0.55/1.5} = P_s \cdot e^{-0.37} = P_s \times 0.69$$

If $P_s = 120$ mmHg, then $P_d \approx 83$ mmHg. This matches the normal diastolic pressure closely, confirming the model.

3.3.3 Solving During Systole

During systole, the left ventricle ejects blood at rate $Q_{in}(t)$. For a simplified rectangular ejection pulse (flow Q_0 for duration T_s):

[Useful Approximation]

$$C\frac{\mathrm{d}P}{\mathrm{d}t} = Q_0 - \frac{P}{R}, \quad 0 \leq t \leq T_s \tag{3.6}$$

The steady-state solution (if ejection continued indefinitely) would be $P_\infty = Q_0 R$. The full solution with initial condition $P(0) = P_d$:

$$P(t) = Q_0 R + (P_d - Q_0 R)\, e^{-t/\tau}, \quad 0 \leq t \leq T_s \tag{3.7}$$

Pressure rises exponentially toward $Q_0 R$ during systole, then falls exponentially toward zero during diastole. The resulting pressure waveform — a rapid rise followed by a slower exponential decay — is qualitatively identical to the measured arterial pressure waveform.

3.4 The Derivation: Why $\tau = RC$ Determines Pulse Pressure

The pulse pressure $PP = P_s - P_d$ is the key clinical output of the Windkessel model. To derive it, note that at the end of diastole (duration T_d), pressure must return to P_d (steady state over many beats). From equation (3.4):

$$P_d = P_s \cdot e^{-T_d/\tau} \tag{3.8}$$

Therefore:

$$P_s = P_d \cdot e^{T_d/\tau} \tag{3.9}$$

And pulse pressure:

$$PP = P_s - P_d = P_d \left(e^{T_d/\tau} - 1 \right) \tag{3.10}$$

Three clinical predictions follow directly:

***Prediction 1 — Stiff arteries increase pulse pressure.** Reduced C reduces $\tau = RC$, making $e^{T_d/\tau}$ larger, increasing PP.*

***Prediction 2 — Tachycardia reduces pulse pressure.** Higher heart rate reduces T_d, reducing $e^{T_d/\tau}$, reducing PP. This is why pulse pressure narrows during tachycardia.*

***Prediction 3 — High resistance increases pulse pressure.** Higher R increases τ, but also increases P_d (from $MAP = CO \times TPR$). The net effect on PP depends on which dominates — captured by the full two-element model.*

Why pulse pressure predicts cardiovascular risk: *A pulse pressure above 60 mmHg is an independent risk factor for cardiovascular events in people over 60. From the Windkessel model, high pulse pressure means low compliance (C) — stiff arteries. Stiff arteries increase the pulsatile load on the left ventricle (it must generate higher peak pressures for the same stroke volume), increase the velocity of the pulse wave, and deliver pulsatile flow to organs (especially the kidney and brain) that are designed for steady flow. The mathematics explains every one of these consequences.*

3.5 The Three-Element Windkessel: Adding Aortic Impedance

The two-element model overestimates early systolic pressure rise because it ignores the resistance of the aorta itself to the sudden ejection of blood. The three-element Windkessel adds a series resistance Z_0 (characteristic impedance of the aorta) to the circuit:

$$P(t) = Z_0 \cdot Q_{in}(t) + \frac{1}{C} \int_0^t Q_C(t')\, \mathrm{d}t' \tag{3.11}$$

where $Q_C = Q_{in} - P/R$ is the flow into the compliance element. The three-element model produces a pressure waveform with a sharp early systolic peak followed by the slower exponential filling and decay — matching the measured aortic pressure waveform with remarkable accuracy.

The characteristic impedance Z_0 is related to the aortic compliance and the Windkessel time constant:

$$Z_0 = \sqrt{\frac{\rho}{\pi^2 R_{ao}^2 C_{ao}}} \tag{3.12}$$

where R_{ao} is the aortic radius and C_{ao} is the aortic wall compliance per unit length. Typical value: $Z_0 \approx 0.06$ mmHg·s/mL for the human aorta.

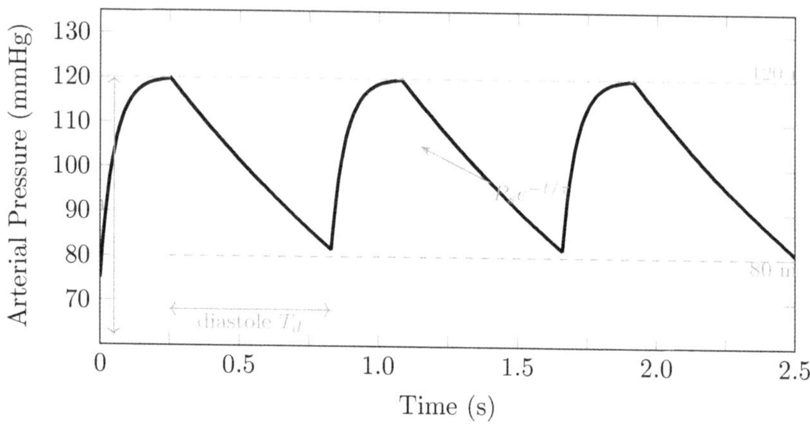

Figure 3.1: Windkessel arterial pressure waveform across three cardiac cycles. During systole, the ventricle ejects blood, raising aortic pressure to 120 mmHg. During diastole, pressure decays exponentially: $P(t) = P_s e^{-t/\tau}$ with $\tau = RC \approx 1.5$ s. Pulse pressure PP $= P_s - P_d \approx 40$ mmHg. As arterial compliance C falls with ageing, τ shortens and PP widens.

3.6 The Clinical Interpretation: Ageing and the Loss of the Windkessel

Arterial stiffening with age is one of the most clinically important and mathematically tractable processes in cardiovascular medicine. As the aortic wall stiffens:

1. *C decreases (wall becomes less compliant)*

2. *$\tau = RC$ decreases*

3. *Pulse pressure increases (equation 3.10)*

4. *Pulse wave velocity increases: $v_{PWV} = \sqrt{1/(\rho C_{wall})}$*

5. *The reflected pulse wave arrives earlier in systole (normally it arrives in diastole, boosting coronary perfusion; now it arrives in systole, increasing afterload)*

The Windkessel model quantifies each of these changes. Pulse wave veloc-

*ity (PWV) is now routinely measured clinically as a non-invasive measure
of arterial stiffness — target < 10 m/s in adults.*

Age	C (mL/mmHg)	τ (s)	PP (mmHg)	PWV (m/s)
25 years	1.8	1.80	38	5.5
45 years	1.3	1.30	45	7.2
65 years	0.8	0.80	58	9.8
85 years	0.4	0.40	82	14.1

*The table shows that aortic compliance halves every 20 years after age 25
in a typical population. Pulse pressure more than doubles from 25 to 85
years. This is not a disease — it is normal ageing of the Windkessel. But
it has the haemodynamic consequence of isolated systolic hypertension in
nearly all very elderly patients.*

3.7   The Worked Example: Estimating Compliance from a Blood Pressure Waveform

*A patient has the following measurements from continuous arterial line
monitoring: systolic pressure 148 mmHg, diastolic pressure 72 mmHg,
heart rate 68 bpm (cycle length $T = 0.882$ s), diastolic duration
$T_d = 0.59$ s (two-thirds of cycle), cardiac output (from thermodilution)
5.2 L/min.*

Step 1 — Compute MAP and TPR.

$$MAP = \frac{148 + 2 \times 72}{3} = \frac{292}{3} \approx 97 \ mmHg$$

$$R = TPR = \frac{MAP}{CO} = \frac{97}{5,200/60} = \frac{97}{86.7} \approx 1.12 \ mmHg \cdot s/mL$$

Step 2 — Estimate compliance from diastolic decay. *From equation (3.8):*

$$P_d = P_s \cdot e^{-T_d/\tau}$$

$$72 = 148 \cdot e^{-0.59/\tau}$$

$$e^{-0.59/\tau} = \frac{72}{148} = 0.486$$

$$-\frac{0.59}{\tau} = \ln(0.486) = -0.721$$

$$\tau = \frac{0.59}{0.721} = 0.818 \ s$$

Step 3 — Compute compliance.

$$C = \frac{\tau}{R} = \frac{0.818}{1.12} \approx 0.73 \ mL/mmHg \quad (range \ 0.61\text{–}0.87 \ from \ \pm 10\% \ measurement \ $$

This compliance estimate of 0.73 ± 0.13 mL/mmHg is below the normal value of 1.5 mL/mmHg, consistent with arterial stiffening. From the table above, this corresponds to a physiological age of approximately 70–75 years — even if the calendar age is younger. Arterial stiffness can be estimated from nothing more than a blood pressure waveform and a cardiac output measurement.

Model Assumptions, Chapter 3. *What the Windkessel model assumes: (1) The entire arterial compliance is concentrated in a single capacitor. (2) Peripheral resistance is uniform and constant. (3) Aortic flow is an external input (not coupled back to ventricular mechanics). Where it breaks down: Wave reflections from arterial branch points (ignored in 2-element Windkessel, partially captured in 3-element). Nonlinear compliance (arterial stiffness increases with pressure, making C pressure-dependent). The diastolic decay is approximately exponential but not perfectly so in measured data. Use the time constant estimate as a population-level index, not a precise individual value.*

3.8 The Software Module

Module 3.1 — Windkessel Model Simulator

Input: *Systolic pressure (mmHg); diastolic pressure (mmHg); heart rate (bpm); cardiac output (L/min); diastolic fraction of* *cycle (default 0.67); optional: measured compliance C for simulation.*

Processing:

1. *Compute MAP and TPR (Module 1.1).*

2. *Estimate τ from the diastolic decay equation (3.8).*

3. *Compute $C = \tau/R$.*

4. *Simulate the two-element Windkessel ODE (equation 3.2) over 3–5 cardiac cycles using scipy.integrate.solve_ivp.*

5. *Compute predicted pulse pressure from equation (3.10).*

6. *Predict change in pulse pressure for given change in C (ageing simulation).*

Output: *Estimated arterial compliance C; time constant τ; simulated pressure waveform (systolic and diastolic phases); pulse pressure; pulse wave velocity estimate; ageing simulation curve (PP vs. C).*

Key equations:

$$C\frac{\mathrm{d}P}{\mathrm{d}t} = Q_{in}(t) - \frac{P}{R}$$

$$P(t) = P_s e^{-t/\tau} \quad (diastole)$$

$$PP = P_d(e^{T_d/\tau} - 1)$$

Full implementation at themathematicsoftheliving-body.com/modules.

3.9 Chapter Summary

1. *Arterial compliance $C = \mathrm{d}V/\mathrm{d}P$ (mL/mmHg) is the ability of arteries to accept stroke volume with limited pressure rise. The aorta acts as a Windkessel — a pressure reservoir that smooths pulsatile*

cardiac output into nearly steady capillary flow.

2. *The two-element Windkessel ODE is* $C\,\mathrm{d}P/\mathrm{d}t = Q_{in} - P/R$. *During diastole, pressure decays exponentially:* $P(t) = P_s e^{-t/\tau}$, *where* $\tau = RC$ *is the diastolic time constant* $(\approx 1.5\ s\ normally)$.

3. *Pulse pressure* $PP = P_d(e^{T_d/\tau} - 1)$. *It increases when compliance decreases (arterial stiffening), when heart rate decreases (longer diastole), or when peripheral resistance increases.*

4. *Compliance can be estimated from the blood pressure waveform alone: fit the diastolic decay to* $P_d = P_s e^{-T_d/\tau}$, *extract* τ, *then* $C = \tau/R$.

5. *Aortic compliance halves approximately every 20 years after age 25. This produces isolated systolic hypertension in the elderly — a direct, mathematically predictable consequence of Windkessel deterioration.*

6. *The three-element Windkessel adds characteristic aortic impedance* Z_0, *improving the fit to the early systolic pressure peak and correctly modelling the effect of aortic valve disease on the pressure waveform.*

Key Equations, Chapter 3

$$C = \frac{\mathrm{d}V}{\mathrm{d}P} \quad \text{(compliance)}$$

$$C\frac{\mathrm{d}P}{\mathrm{d}t} = Q_{in} - \frac{P}{R} \quad \text{(Windkessel ODE)}$$

$$P(t) = P_s e^{-t/\tau}, \quad \tau = RC \quad \text{(diastolic decay)}$$

$$PP = P_d\left(e^{T_d/\tau} - 1\right) \quad \text{(pulse pressure)}$$

Key References, Chapter 3

1. Westerhof, N. et al. (2009). *The arterial Windkessel. Med. Biol. Eng. Comput.* 47, 131–141. [Comprehensive review of 2- and 3-element Windkessel models.]

2. Nichols, W.W. & O'Rourke, M.F. (2011). McDonald's Blood Flow in Arteries, 6th ed. Hodder Arnold. [Standard reference for arterial compliance and pulse wave analysis.]

3. Laurent, S. et al. (2006). Aortic stiffness is an independent predictor of all-cause and cardiovascular mortality in hypertensive patients. Hypertension 37, 1236–1241. [Clinical evidence linking PWV to mortality.]

Chapter 4

The Pressure-Volume Loop: Cardiac Work as an Integral

The pressure-volume loop is the electrocardiogram of the haemodynamicist. One loop tells you the preload, the afterload, the contractility, the stroke work, and the efficiency of the heart — all at once, from first principles.

— Zachariah Sinkala

4.1 The Physiology: One Heartbeat, Four Phases

A single heartbeat consists of four distinct mechanical phases. Understanding each phase is essential before the mathematics can be written down, because the pressure-volume loop is simply these four phases traced sequentially in pressure-volume space.

Phase 1 — Isovolumetric contraction. *The mitral valve has just closed (end of diastole). The ventricle begins to contract but neither valve is open — pressure rises rapidly while volume stays constant. This phase ends when left ventricular pressure exceeds aortic pressure and the aortic valve opens.*

Phase 2 — Ejection. *The aortic valve is open. Blood flows into the aorta as the ventricle continues to contract. Volume falls (stroke volume is ejected) while pressure continues to rise to its peak (systolic pressure) and then begins to fall as the ventricle starts to relax. The aortic valve closes when LV pressure falls below aortic pressure.*

Phase 3 — Isovolumetric relaxation. *Both valves are closed again. The ventricle relaxes rapidly — pressure falls while volume stays constant. This phase ends when LV pressure falls below left atrial pressure and the mitral valve opens.*

Phase 4 — Filling (diastole). *The mitral valve is open. Blood flows from the left atrium into the ventricle. Volume increases while pressure rises only slightly (because the relaxed ventricle is compliant). This phase ends when the next systole begins.*

Tracing these four phases in a pressure (P) versus volume (V) diagram produces a closed loop — the pressure-volume loop. The area enclosed by the loop is the stroke work performed by the heart in one beat.

4.2 The Observation: What the Loop Looks Like

In a normal left ventricle, the PV loop has a characteristic rectangular shape with rounded corners:

- ***Bottom edge*** *(diastolic filling): Volume increases from ≈ 50 mL (end-systolic volume, ESV) to ≈ 120 mL (end-diastolic volume, EDV) along the passive filling curve. Pressure rises from ≈ 5 to ≈ 12 mmHg.*

- **Left edge** *(isovolumetric contraction): Pressure rises steeply from ≈ 12 to ≈ 80 mmHg at constant volume (EDV ≈ 120 mL).*

- **Top edge** *(ejection): Volume falls from 120 to 50 mL as pressure rises to peak systolic (≈ 120 mmHg) then falls back to ≈ 80 mmHg.*

- **Right edge** *(isovolumetric relaxation): Pressure falls from ≈ 80 to ≈ 5 mmHg at constant volume (ESV).*

Stroke volume $SV = EDV - ESV = 120 - 50 = 70$ mL. Ejection fraction $EF = SV/EDV = 70/120 \approx 58\%$. Normal EF is $\geq 55\%$.

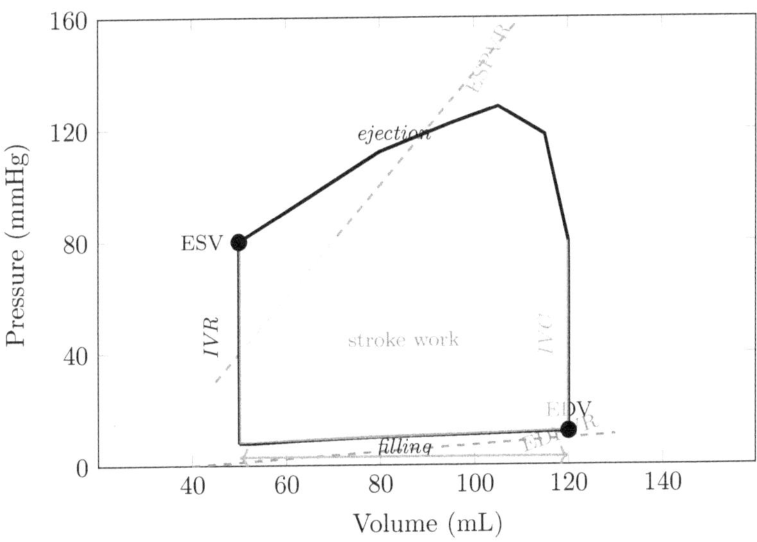

Figure 4.1: The normal left ventricular pressure-volume loop. Diastolic filling (bottom edge) follows the EDPVR (blue dashed). Isovolumetric contraction (IVC, left vertical edge) raises pressure at constant volume. Ejection (top edge) reduces volume as pressure peaks. Isovolumetric relaxation (IVR, right vertical edge) restores diastolic pressure. The shaded area is the stroke work $W = \oint P \, dV$. The ESPVR slope E_{es} (red dashed) defines contractility.

4.3   The Mathematics: Two Fundamental Lines

The PV loop is bounded by two linear relationships that define the limits of ventricular performance: the end-diastolic pressure-volume relationship (EDPVR) and the end-systolic pressure-volume relationship (ESPVR).

4.3.1   End-Diastolic Pressure-Volume Relationship

The EDPVR describes how pressure rises as the ventricle fills during diastole. For the normal ventricle, it is approximately exponential over the physiological range:

$$P_{ED}(V) = \alpha \left(e^{\beta(V - V_0)} - 1 \right) \tag{4.1}$$

where α and β are constants describing wall stiffness and V_0 is the unstressed volume. Typical values for the left ventricle: $\alpha \approx 0.5$ mmHg, $\beta \approx 0.04$ mL^{-1}, $V_0 \approx 20$ mL. Over the normal filling range (50–130 mL), the EDPVR is nearly linear with a slope (diastolic stiffness) of approximately 0.1 mmHg/mL.

4.3.2   End-Systolic Pressure-Volume Relationship

The ESPVR is the most important relationship in cardiac physiology. It describes the maximum pressure the ventricle can generate at any given volume during systole. For the normal ventricle it is linear:

[Useful Approximation]

$$P_{ES}(V) = E_{es}(V - V_d) \tag{4.2}$$

where E_{es} is the end-systolic elastance *(mmHg/mL) and V_d is the volume-axis intercept (the volume at which the ventricle would generate zero pressure).*

E_{es} is the quantitative measure of myocardial contractility *— the intrinsic force-generating capacity of the ventricle, independent of preload and*

afterload. Typical population values: $E_{es} \approx 1.5\text{--}2.5$ mmHg/mL for the left ventricle (range reflects age, sex, and loading conditions).

The ESPVR line has a crucial property: the end-systolic point of every PV loop, regardless of preload or afterload, lies on the same ESPVR line (as long as contractility has not changed). This makes E_{es} a load-independent measure of cardiac function.

4.3.3   Stroke Work as the Area of the Loop

The mechanical work done by the ventricle in one beat is the integral of pressure over volume change:

[Core Principle]

$$W_{stroke} = \oint P \, dV \tag{4.3}$$

This closed integral — the area enclosed by the PV loop — is the stroke work in units of mmHg·mL. To convert to Joules: 1 mmHg·mL $= 1.333 \times 10^{-4}$ J.

For the normal ventricle approximated as a rectangle:

$$W_{stroke} \approx SV \times MAP = 70 \times 93 = 6{,}510 \ mmHg \cdot mL \approx 0.87 \ J \tag{4.4}$$

Cardiac power (work per unit time):

$$\dot{W} = W_{stroke} \times HR = 0.87 \times 72/60 \approx 1.04 \ W \tag{4.5}$$

The resting heart produces approximately 1 watt of mechanical power — remarkably, the same as a moderately bright LED. During maximal exercise, cardiac power can reach 10–20 W.

4.4   The Derivation: Preload, Afterload, and Contractility from First Principles

The three determinants of cardiac output — preload, afterload, and contractility — are defined precisely by the PV loop.

4.4.1   Preload: The Frank-Starling Effect

Preload is the end-diastolic volume (EDV) — the volume in the ventricle just before systole begins. Increasing preload moves the starting point of the PV loop to the right along the EDPVR.

If the ESPVR slope (E_{es}) remains constant (contractility unchanged), the end-systolic point also moves along the ESPVR. Stroke volume increases because:

$$SV = EDV - ESV = EDV - \left(V_d + \frac{P_{AO}}{E_{es}} \right) \tag{4.6}$$

where P_{AO} is aortic pressure (afterload). Increased EDV increases SV — this is the Frank-Starling mechanism, derived directly from the geometry of the PV loop. Chapter 5 derives the Frank-Starling curve from the sarcomere length-tension relationship.

4.4.2   Afterload: Aortic Pressure

Afterload is the pressure the ventricle must overcome to eject blood — approximated by aortic pressure P_{AO}. From equation (4.2), at the end-systolic point:

$$P_{AO} = E_{es}(ESV - V_d)$$

Therefore:

$$ESV = V_d + \frac{P_{AO}}{E_{es}} \tag{4.7}$$

Higher afterload (higher P_{AO}) increases ESV, reducing SV (for the same EDV). The PV loop becomes taller and narrower — the hallmark of pressure overload, as seen in hypertension and aortic stenosis.

4.4.3   Contractility: The ESPVR Slope

Contractility is E_{es} — the slope of the ESPVR. When contractility increases (e.g. with inotropes like dobutamine), the ESPVR rotates upward (steeper slope), the end-systolic point moves up and to the left, ESV decreases, and SV increases. When contractility decreases (e.g. in heart *failure), the ESPVR rotates downward, ESV increases, and SV decreases.*

The key insight: for a given preload (EDV) and afterload (P_{AO}), the stroke volume is completely determined by E_{es}:

$$SV = EDV - V_d - \frac{P_{AO}}{E_{es}} \tag{4.8}$$

This is the fundamental equation of ventricular performance. It reduces the entire complex physiology of cardiac output to three parameters: EDV (preload), P_{AO} (afterload), and E_{es} (contractility).

4.5 The Clinical Interpretation: Reading the PV Loop at the Bedside

Four pathological PV loops tell four clinical stories:

Heart failure with reduced EF (HFrEF). *E_{es} is reduced (ESPVR shifts rightward and flattens). ESV is high, SV is low, EF is low ($< 40\%$). The loop is wide but short. Treatment: increase E_{es} (inotropes) or reduce P_{AO} (vasodilators, ACE inhibitors).*

Hypertension. *P_{AO} is elevated (increased afterload). ESV is high, SV is reduced. The loop is tall and narrow. The heart compensates by hypertrophy (increasing E_{es}), at the cost of reduced compliance. Treatment: reduce P_{AO} (antihypertensives).*

Aortic stenosis. *P_{AO} is effectively increased (ventricle must generate supranormal pressure to open the stenotic valve). The loop is very tall, very narrow, with high peak pressure but reduced SV. The classic "pressure-overloaded" loop.*

Mitral regurgitation. *During systole, some blood leaks backward through the incompetent mitral valve into the low-pressure left atrium. Effective forward SV is reduced despite normal or even elevated total SV. The loop is wide (high ESV and EDV) but the effective stroke work is reduced. The ventricle dilates (increased EDV) to compensate via Frank-Starling.*

> ***The single most useful number in the PV loop:*** *E_{es}, the end-systolic elastance, is the only load-independent measure of myocardial contractility. EF (ejection fraction) depends on both contractility and afterload — a ventricle with normal E_{es} but high afterload will have a reduced EF. E_{es} does not. It is the true measure of the heart's intrinsic pumping capacity, and it is derived from the slope of the end-systolic pressure-volume relationship.*

4.6 Mechanical Efficiency: How Much Work Is Wasted

Not all the energy the heart expends goes into stroke work. Some goes into isovolumetric contraction and relaxation (pressure generated without volume change — pure energy expenditure with no external work done). Define:

Potential energy (PE): *The area between the ESPVR and EDPVR lines, to the left of the PV loop. This is the energy stored in the myocardium at end-systole that is dissipated as heat during isovolumetric relaxation.*

Pressure-volume area (PVA): *The total mechanical energy = stroke work + potential energy:*

$$PVA = W_{stroke} + PE \tag{4.9}$$

Myocardial oxygen consumption $\dot{V}O_2$ is linearly related to PVA:

$$\dot{V}O_2 = a \cdot PVA + b \tag{4.10}$$

where $a \approx 1.5 \times 10^{-5}$ mL O_2/ (mmHg·mL) and b is the basal oxygen consumption for non-mechanical work (ion pumping, maintenance). Mechanical efficiency is:

$$\eta = \frac{W_{stroke}}{E_{mech}} = \frac{W_{stroke}}{PVA} \tag{4.11}$$

Normal mechanical efficiency is approximately 25–35%. The remaining 65–75% of energy is dissipated as heat — the reason the heart is a signif-icant source of body heat.

4.7 The Worked Example: Quantifying Heart Failure from a PV Loop

A patient with dilated cardiomyopathy undergoes pressure-volume loop analysis using conductance catheterisation. Measurements: $EDV = 240$ mL, $ESV = 185$ mL, peak LV pressure $= 95$ mmHg, end-diastolic pressure $= 22$ mmHg, aortic pressure $= 82$ mmHg. Three PV loops at different preloads give ESPVR: $P_{ES} = 0.6(V - 30)$.

Step 1 — Basic haemodynamics.

$$SV = 240 - 185 = 55 \ mL$$
$$EF = 55/240 = 23\% \quad (severely \ reduced, \ normal \geq 55\%)$$

Step 2 — Contractility.

$$E_{es} = 0.6 \ mmHg/mL \quad (normal \approx 2.0 \ mmHg/mL)$$

Contractility is 70% below normal. This is severe systolic dysfunction.

Step 3 — Stroke work.

$$W_{stroke} \approx SV \times MAP = 55 \times 82 = 4{,}510 \ mmHg \cdot mL \approx 0.60 \ J$$

Step 4 — Predicted response to dobutamine. *If dobutamine in-creases E_{es} to 1.2 mmHg/mL (doubled), from equation (4.8):*

$$ESV_{new} = 30 + \frac{82}{1.2} = 30 + 68.3 = 98.3 \ mL$$

$$SV_{new} = 240 - 98.3 = 141.7 \ mL \quad (157\% \ increase!)$$

Doubling E_{es} more than doubles SV — because the failing ventricle was so far from its ESPVR that even a modest improvement in contractility produces a large increase in stroke volume. This is why inotropes are effective in acutely decompensated heart failure even when they have limited long-term benefit.

Model Assumptions, Chapter 4. *What the PV loop model assumes: (1) The ESPVR is linear (it deviates from linearity at very low and very high volumes). (2) E_{es} is load-independent (approximately true over the physiological range; fails in severe ischaemia and at extreme volumes). (3) The ventricle is spatially uniform (no regional wall motion abnormalities). Where it breaks down: After myocardial infarction, regional dysfunction violates the uniform ventricle assumption. The single-beat E_{es} estimate requires the assumption $V_d \approx 30 \ mL$, which adds $\pm 0.3 \ mmHg/mL$ uncertainty to the result.*

4.8   The Software Module

Module 4.1 — Pressure-Volume Loop Simulator

Input: *EDV (mL); ESV (mL); peak LV pressure (mmHg); end-diastolic pressure (mmHg); aortic pressure (mmHg); heart rate (bpm); ESPVR slope E_{es} and intercept V_d (if available from multi-beat data); intervention: change in E_{es}, EDV, or P_{AO}.*

Processing:

1. *Compute SV, EF, stroke work, cardiac power.*

2. *Fit ESPVR from multi-beat data if provided; extract E_{es} and V_d.*

3. *Simulate PV loop for current state.*

4. *Apply intervention: recompute ESV, SV, stroke work using*

equation (4.8).

5. *Compute mechanical efficiency (equation 4.11).*

6. *Classify loop into one of four pathological patterns (HFrEF, hypertension, aortic stenosis, mitral regurgitation).*

Output: *PV loop plot; SV, EF, E_{es}, stroke work, cardiac power; pre- and post-intervention comparison; efficiency; clinical classification.*

Key equations:

$$SV = EDV - V_d - P_{AO}/E_{es}$$

$$W = \oint P \, dV \approx SV \times MAP$$

$$\eta = W_{stroke}/PVA$$

Full implementation at themathematicsoftheliving-body.com/modules.

4.9   Chapter Summary

1. *The cardiac cycle has four phases: isovolumetric contraction, ejection, isovolumetric relaxation, and diastolic filling. Tracing these in pressure-volume space produces the PV loop.*

2. *The ESPVR ($P_{ES} = E_{es}(V - V_d)$) is a load-independent measure of contractility. Its slope E_{es} (normal ≈ 2.0 mmHg/mL) quantifies the heart's intrinsic pumping capacity.*

3. *Stroke volume is determined by three parameters: $SV = EDV - V_d - P_{AO}/E_{es}$. Preload (EDV), afterload (P_{AO}), and contractility (E_{es}) are all captured in one equation.*

4. *Stroke work equals the area of the PV loop: $W \approx SV \times MAP$. Cardiac power at rest ≈ 1 W. Mechanical efficiency is 25–35%.*

5. *In the heart failure worked example, E_{es} was 70% below normal (0.6 vs 2.0 mmHg/mL). Doubling E_{es} with dobutamine increased predicted SV by 157% — demonstrating why inotropes have such dramatic acute effects in decompensated heart failure.*

6. *EF alone is insufficient to characterise ventricular function. E_{es} is load-independent; EF is not. A patient with normal E_{es} but high afterload will have reduced EF despite normal contractility.*

Key Equations, Chapter 4

$$P_{ES}(V) = E_{es}(V - V_d) \quad (ESPVR)$$

$$SV = EDV - V_d - \frac{P_{AO}}{E_{es}} \quad (stroke\ volume)$$

$$W_{stroke} = \oint P\,\mathrm{d}V \approx SV \times MAP$$

$$\eta = \frac{W_{stroke}}{PVA} \quad (mechanical\ efficiency)$$

Key References, Chapter 4

1. *Suga, H. & Sagawa, K. (1974). Instantaneous pressure-volume relationships and their ratio in the excised, supported canine left ventricle. Circ. Res. 35, 117–126. [Original derivation of E_{es} as a load-independent index of contractility.]*

2. *Starling, M.R. (1993). Left ventricular end-systolic pressure-volume relations: comparison of theoretical models with experimental data. J. Am. Coll. Cardiol. 22, 460–469.*

3. *Nozawa, T. et al. (1988). Efficiency of energy transfer from pressure-volume area to external mechanical work increases with contractile state and heart rate in the right ventricle. Circulation 77, 1116–1124. [Pressure-volume area and mechanical efficiency.]*

Chapter 5

Frank-Starling: Deriving the Curve from First Principles

The heart pumps more when it is filled more. Every medical student knows this. Almost none of them know why — which means almost none of them know when it stops being true, or why the failing heart loses it.

— Zachariah Sinkala

5.1 The Physiology: The Heart Adapts to Its Filling

In 1914, Ernest Starling demonstrated that an isolated heart-lung preparation pumped more blood when venous return was increased — the heart's output automatically matched its input without any nervous system in-

volvement. Otto Frank had described similar results a decade earlier in isolated frog hearts. The Frank-Starling law states: within physiological limits, the force of ventricular contraction is proportional to the end-diastolic fibre length.

In clinical terms: the more blood that fills the heart during diastole (higher preload), the more forcefully it contracts and the greater the stroke volume ejected.

This law is not a curiosity of isolated heart preparations. It is the fundamental mechanism by which:

- *The two ventricles maintain equal output beat by beat (if the right ventricle ejects slightly more, the left ventricle fills slightly more and matches it automatically)*

- *Cardiac output increases during exercise before the nervous system has time to respond*

- *The heart responds to sudden increases in venous return (standing up, fluid bolus) without needing a signal from the brain*

The question this chapter answers is: why *does stretching a muscle fibre make it contract more forcefully? The answer lies at the level of the sarcomere — the basic contractile unit of the myocyte — and the mathematics of the length-tension relationship.*

Before the equations: the molecular machinery that makes the Frank-Starling law possible.

The sarcomere is the contractile unit of cardiac muscle, bounded by two Z-discs and containing interdigitated thick (myosin) and thin (actin) filaments. The Frank-Starling law exists because the force a sarcomere generates depends on how far the filaments overlap — which depends on sarcomere length — which depends on end-diastolic volume.

Troponin and calcium cooperativity. *Actin filaments are coated by tropomyosin, a coiled-coil protein that blocks the myosin-binding* *sites on actin at rest. The troponin complex (TnC, TnI, TnT) acts as the calcium switch. When intracellular Ca^{2+} rises during the action potential plateau, Ca^{2+} binds to TnC (cardiac isoform: TNNC1), causing a conformational change that moves tropomyosin away from the myosin-binding sites, permitting cross-bridge cycling.*

Length-dependent calcium sensitivity. *The key molecular mechanism underlying Frank-Starling is that TnC's affinity for Ca^{2+} increases when the sarcomere is stretched. Two mechanisms contribute: (1) reduced inter-filament spacing at longer sarcomere lengths increases electrostatic cross-bridge formation probability; (2) stretch-induced conformational changes in TnI reduce its inhibitory interaction with TnC, increasing Ca^{2+} binding affinity. The net result: at longer sarcomere length (higher preload), the same Ca^{2+} transient activates more cross-bridges — the cellular basis of Frank-Starling that the Hill equation in Section 5.3 captures mathematically.*

Titin: the molecular spring. *The giant elastic protein titin (encoded by TTN, the largest human gene at 364 exons) spans from the Z-disc to the M-line, acting as a molecular spring that generates passive restoring force when the sarcomere is stretched beyond its slack length. Titin stiffness sets the diastolic compliance and the slope of the EDPVR. Mutations in TTN are the most common genetic cause of dilated cardiomyopathy — a disease of reduced sarcomere compliance and impaired Frank-Starling reserve.*

5.2 The Observation: The Length-Tension Relationship

The fundamental experiment is simple. Take an isolated papillary muscle. Stretch it to different lengths. Stimulate it electrically. Measure the peak force (tension) it generates. Plot force versus resting length. The result is the length-tension curve*:*

- *At very short lengths (sarcomere length < 1.6 μm): force is low.*

Thick and thin filaments overlap excessively, interfering with cross-bridge cycling.

- *At optimal length (sarcomere length ≈ 2.2 μm): force is maximal. Overlap between thick and thin filaments is optimal — maximum number of cross-bridges can form.*

- *At longer lengths (sarcomere length > 2.2 μm): force initially remains near maximal then falls as filaments are pulled apart and fewer cross-bridges can form.*

- *At very long lengths (> 3.6 μm): force drops to zero. Thin and thick filaments no longer overlap.*

The cardiac muscle in the intact ventricle normally operates at sarcomere lengths of 1.8–2.2 μm — on the ascending limb of the length-tension curve. Increasing end-diastolic volume stretches the sarcomeres from 1.8 toward 2.2 μm, increasing force production. This is the cellular basis of the Frank-Starling law.

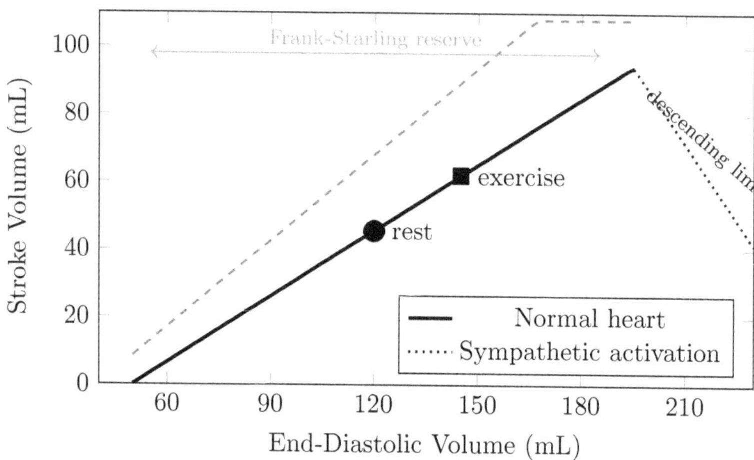

Figure 5.1: The Frank-Starling curve. Stroke volume rises steeply on the ascending limb as EDV increases (sarcomere length approaches optimal 2.2 μm). The plateau represents maximal filament overlap. The descending limb (dotted) is rarely reached in the intact circulation. Sympathetic activation (blue dashed) shifts the entire curve upward by increasing E_{es} — independent of preload.

5.3 The Mathematics: From Sarcomere to Ventricle

5.3.1 The Cross-Bridge Model

The molecular basis of muscle force generation is the cross-bridge cycle: myosin heads attach to actin filaments, rotate (power stroke), detach, and reattach further along the filament. The number of cross-bridges that can form simultaneously depends on the overlap between thick (myosin) and thin (actin) filaments.

Define the overlap fraction $\phi(L_s)$ as the fraction of available myosin heads that can bind to actin at sarcomere length L_s:

[Useful Approximation]

$$\phi(L_s) = \begin{cases} 0 & L_s < 1.27 \ \mu m \\[2mm] \dfrac{L_s - 1.27}{0.93} & 1.27 \leq L_s \leq 2.20 \ \mu m \\[2mm] 1 & 2.20 \leq L_s \leq 2.25 \ \mu m \\[2mm] \dfrac{3.65 - L_s}{1.40} & 2.25 < L_s \leq 3.65 \ \mu m \\[2mm] 0 & L_s > 3.65 \ \mu m \end{cases} \tag{5.1}$$

The active force generated by the sarcomere is proportional to the overlap fraction and the maximum isometric force $F_{\max}$:

$$F_{active}(L_s) = F_{\max} \cdot \phi(L_s) \tag{5.2}$$

5.3.2 Calcium and the Cooperativity Effect

A critical additional mechanism amplifies the Frank-Starling effect beyond what filament overlap alone predicts: calcium sensitivity increases with sarcomere length.

When a myocyte is stretched, the troponin-tropomyosin regulatory complex becomes more sensitive to calcium. This means that at longer sarcomere lengths, the same intracellular calcium concentration activates more cross-bridges. The Hill equation captures this:

$$F_{active}(L_s, [Ca^{2+}]) \;=\; F_{\max} \cdot \phi(L_s) \cdot \frac{[Ca^{2+}]^n}{K_{1/2}(L_s)^n + [Ca^{2+}]^n} \tag{5.3}$$

where $n \approx 3\text{--}4$ is the Hill cooperativity coefficient and $K_{1/2}(L_s)$ is the calcium concentration for half-maximal activation, which decreases with increasing sarcomere length:

$$K_{1/2}(L_s) \;=\; K_{1/2}^0 \cdot e^{-\kappa(L_s - L_s^0)} \tag{5.4}$$

with $\kappa \approx 0.8$ μm^{-1} and $L_s^0 \approx 1.9$ μm. As L_s increases, $K_{1/2}$ decreases, the Hill curve shifts left, and more force is generated at the same $[Ca^{2+}]$. This calcium-sensitisation accounts for approximately 30–50% of the Frank-Starling effect in the intact heart.

5.3.3   From Sarcomere Force to Ventricular Pressure

To connect sarcomere mechanics to the ventricular pressure-volume relationship, use the law of Laplace for a thick-walled sphere (approximation for the ventricle):

[Useful Approximation]

$$P = \frac{2h\sigma}{r} \tag{5.5}$$

where P is ventricular pressure, h is wall thickness, σ is myocardial stress (force per unit area), and r is the ventricular radius. Assuming wall volume is constant as the ventricle fills ($h \cdot r^2 = const$), and relating radius to volume ($V = \frac{4}{3}\pi r^3$):

$$r = \left(\frac{3V}{4\pi} \right)^{1/3} \tag{5.6}$$

Sarcomere length relates to ventricular radius by the fibre strain:

$$L_s = L_{s,0} \left(\frac{r}{r_0}\right)^{2/3} \tag{5.7}$$

where $L_{s,0}$ and r_0 are the reference sarcomere length and radius at zero transmural pressure. Combining equations (5.2), (5.5), and (5.7) gives the Frank-Starling relationship between EDV and peak systolic pressure — the ESPVR of Chapter 4 emerges from sarcomere mechanics.

5.4 The Derivation: The Frank-Starling Curve as an ODE

The stroke volume as a function of preload follows from combining the sarcomere model with the ventricular mechanics of Chapter 4:

$$SV(EDV) = EDV - V_d - \frac{P_{AO}}{E_{es}(EDV)} \tag{5.8}$$

where $E_{es}(EDV)$ now depends on preload (through the sarcomere length-tension relationship):

$$E_{es}(EDV) = E_{es}^0 \cdot \phi(L_s[EDV]) \cdot \frac{[Ca^{2+}]^n}{K_{1/2}(L_s[EDV])^n + [Ca^{2+}]^n} \tag{5.9}$$

This is the Frank-Starling curve written as a function of end-diastolic volume. On the ascending limb ($L_s < 2.2 \ \mu m$): ϕ increases with EDV, E_{es} increases, SV increases. On the descending limb ($L_s > 2.2 \ \mu m$): ϕ decreases, E_{es} decreases, SV decreases.

The descending limb — where stretching the heart reduces output — is rarely reached in the intact circulation because venous capacitance limits filling before sarcomere lengths become excessive. It is seen in acute right heart failure (massive PE) where the right ventricle becomes grossly dilated.

5.5 The Clinical Interpretation: When Frank-Starling Saves You and When It Fails You

5.5.1 The Normal Heart: Using the Ascending Limb

In a healthy person standing up, venous return briefly falls (gravity pools blood in the legs). EDV falls slightly. The Frank-Starling mechanism reduces SV proportionally — but simultaneously, the baroreflex (Chapter 9) increases heart rate and the autonomic system increases contractility. The net result: cardiac output is maintained.

During exercise, venous return increases (muscle pump, respiratory pump). EDV increases. The Frank-Starling mechanism increases SV even before the heart rate and contractility increases driven by the sympathetic nervous system have fully activated. This is why cardiac output begins rising within the first few seconds of exercise.

5.5.2 The Failing Heart: Loss of Frank-Starling Reserve

In heart failure with reduced ejection fraction (HFrEF), the ESPVR is shifted rightward and flattened (E_{es} is reduced). The ventricle dilates chronically (EDV increases) — the body's attempt to use the Frank-Starling mechanism to maintain stroke volume.

But the Frank-Starling reserve is finite. Once EDV is elevated to 200–250 mL, sarcomere lengths approach 2.2 μm and further increases in EDV produce little additional SV increase. The Frank-Starling curve is flatter in the failing heart: the slope $dSV/dEDV$ is reduced. In severe heart failure, the curve becomes essentially flat — the heart can no longer augment output in response to increased filling.

This explains why heart failure patients are exquisitely sensitive to volume: too little fluid reduces EDV below the flat portion of the curve (no reserve), while too much fluid floods the pulmonary circulation (elevated filling pressures) without increasing output.

The clinical rule from the Frank-Starling mathematics: *In the normal heart, $\mathrm{d}SV/\mathrm{d}EDV \approx 0.6$ mL/mL on the ascending limb — a 10 mL increase in EDV increases SV by 6 mL. In severe heart failure, this slope falls to ≈ 0.1 mL/mL. A fluid challenge that would increase CO by 600 mL/min in a normal heart produces only 100 mL/min in severe heart failure — and the elevated filling pressure it causes may worsen pulmonary oedema. This is why fluid challenges must be interpreted differently in the failing heart.*

5.6 Starling Curves in Clinical Practice: The Fluid Responsiveness Problem

One of the most common questions in critical care is: will this patient's cardiac output increase if I give them a fluid bolus? *This is the fluid responsiveness question, and it is a direct clinical application of the Frank-Starling curve.*

A patient is fluid-responsive if they are on the ascending limb of their Frank-Starling curve — where increasing EDV increases SV. They are non-responsive if they are at the plateau.

The passive leg raise (PLR) test autotransfuses approximately 300 mL from the legs to the central circulation, temporarily increasing EDV. If SV increases by $\geq 10\%$ (measured by pulse contour analysis or echocardiography), the patient is fluid-responsive. The test is reversible — lowering the legs returns the blood to the legs.

Mathematically, the PLR tests the local slope of the Frank-Starling curve at the patient's current operating point:

$$\text{Fluid responsive} \iff \left. \frac{\mathrm{d}SV}{\mathrm{d}EDV} \right|_{current} > 0.10 \ mL/mL \qquad (5.10)$$

Static measures (CVP, PCWP) tell you the current EDV, not where you are on the curve. Dynamic measures (PLR, pulse pressure variation) tell you the local slope — which is what actually matters for the treatment decision.

5.7 The Worked Example: Predicting Fluid Responsiveness

A 54-year-old man in septic shock has the following measurements: HR 112 bpm, MAP 62 mmHg, CVP 8 mmHg. Stroke volume index (SVI) by echocardiography 35 mL/m² (normal ≥ 40). Passive leg raise increases SVI to 42 mL/m².

Step 1 — Assess fluid responsiveness.

$$\Delta SVI = \frac{42 - 35}{35} \times 100\% = 20\% \geq 10\% \quad \Rightarrow \quad \textit{fluid responsive}$$

The patient is on the ascending limb of the Frank-Starling curve.

Step 2 — Estimate the local Frank-Starling slope. *The PLR added approximately 300 mL to the central circulation. EDV increased by approximately $\Delta EDV \approx 300 \times 0.5 = 150$ mL (assuming 50% goes to the ventricle). SV increased by $\Delta SV = 7$ mL/m².*

$$\left.\frac{\mathrm{d}SV}{\mathrm{d}EDV}\right|_{est} \approx \frac{7}{150} = 0.047 \; mL/mL$$

This is a relatively shallow slope, suggesting the patient is approaching the plateau of their Frank-Starling curve.

Step 3 — Treatment decision. *Give a 500 mL fluid bolus. Predicted change in SV:*

$$\Delta SV_{predicted} \approx 0.047 \times 500 \times 0.5 \approx 11.75 \; mL$$

Predicted change in CO:

$$\Delta CO = \Delta SV \times HR = 0.012 \times 112 \approx 1.3 \; L/min$$

A 1.3 L/min increase in CO is clinically significant. Administer the bolus *and reassess SVI immediately after. If the slope flattens further (smaller ΔSVI per bolus), the plateau has been reached and further fluids will increase filling pressure without benefiting cardiac output.*

Model Assumptions, Chapter 5. *What the Frank-Starling model assumes: (1) The overlap fraction $\phi(L_s)$ is uniform across all sarcomeres. (2) Calcium sensitivity increases monotonically with sarcomere length (true over physiological range, breaks down at extreme stretch). (3) The Laplace approximation treats the ventricle as a sphere. Where it breaks down: Sarcomere length is not uniform across the ventricular wall (subendocardial sarcomeres are shorter than subepicardial). Titin mutations alter passive stiffness in ways not captured by the Hill equation alone. The descending limb of the length-tension curve is rarely reached in the intact heart.*

5.8   The Software Module

Module 5.1 — Frank-Starling Curve Simulator and Fluid Responsiveness Predictor

Input: *Current EDV or SVI; current SV or CO; passive leg raise ΔSV (if available); E_{es} and V_d from Module 4.1; aortic pressure; proposed fluid bolus volume.*

Processing:

1. *Compute the Frank-Starling curve (SV vs. EDV) using equation (5.8) over EDV range 50–300 mL.*

2. *Mark the current operating point.*

3. *Compute local slope $\mathrm{d}SV/\mathrm{d}EDV$ at the operating point.*

4. *Classify as fluid responsive (> 0.10) or non-responsive.*

5. *Predict SV and CO after proposed fluid bolus using the local slope.*

6. *Simulate the failing heart: shift ESPVR to reduced E_{es} and*

show the flattened curve.

Output: *Frank-Starling curve plot with operating point; fluid responsiveness classification; predicted post-bolus CO; comparison of normal vs. failing heart curves.*

Key equations:

$$SV = EDV - V_d - P_{AO}/E_{es}(EDV)$$

$$\frac{\mathrm{d}SV}{\mathrm{d}EDV} > 0.10 \Leftrightarrow \textit{fluid responsive}$$

Full implementation at themathematicsoftheliving-body.com/modules.

5.9   Chapter Summary

1. *The Frank-Starling law arises from two molecular mechanisms: increased filament overlap (quantified by $\phi(L_s)$) and increased calcium sensitivity ($K_{1/2}$ decreases with sarcomere length). Both increase active force when the myocyte is stretched.*

2. *The Frank-Starling curve (SV vs. EDV) has an ascending limb (operating range of the normal heart, $L_s < 2.2~\mu m$), a plateau, and a descending limb (rarely reached in vivo).*

3. *Heart failure reduces the slope of the Frank-Starling curve: from $\approx$ 0.6 mL/mL normally to $\approx$ 0.1 mL/mL in severe HFrEF. This makes the failing heart insensitive to volume loading and sensitive to volume overload.*

4. *Fluid responsiveness is the clinical application: a patient is fluid responsive if their local Frank-Starling slope exceeds 0.10 mL/mL. The PLR test measures this slope non-invasively.*

5. *Static filling pressures (CVP, PCWP) measure position on the volume axis, not slope. Dynamic tests (PLR, pulse pressure variation) measure slope — the quantity that actually predicts the response to*

fluid.

6. *In the worked example, a 20% SVI increase with PLR confirmed fluid responsiveness, and the estimated local slope predicted a 1.3 L/min CO increase from a 500 mL bolus.*

Key Equations, Chapter 5

$$F_{active} = F_{\max} \cdot \phi(L_s) \quad \text{(length-tension)}$$

$$F_{active} = F_{\max} \cdot \phi(L_s) \cdot \frac{[Ca^{2+}]^n}{K_{1/2}(L_s)^n + [Ca^{2+}]^n}$$

$$P = \frac{2h\sigma}{r} \quad \text{(Laplace, ventricle)}$$

$$\text{Fluid resp.} \Leftrightarrow \frac{\mathrm{d}SV}{\mathrm{d}EDV} > 0.10 \ mL/mL$$

Key References, Chapter 5

1. *Frank, O. (1895). Zur Dynamik des Herzmuskels. Z. Biol. 32, 370–447. [Original description of the length-force relationship.]*

2. *Allen, D.G. & Kentish, J.C. (1985). The cellular basis of the length-tension relation in cardiac muscle. J. Mol. Cell. Cardiol. 17, 821–840. [Calcium cooperativity and length-dependent activation.]*

3. *Monnet, X. et al. (2006). Passive leg raising predicts fluid responsiveness in the critically ill. Crit. Care Med. 34, 1402–1407. [Clinical validation of the PLR test for Frank-Starling assessment.]*

Chapter 6

Cardiac Output: The Fick Principle and Beyond

Cardiac output cannot be measured directly. It must be inferred — from what the body consumes, from the dilution of an indicator, from the shape of a pressure waveform. Each method is a different application of the same conservation law.

— Zachariah Sinkala

6.1 The Physiology: What Cardiac Output Means

Cardiac output (CO) is the volume of blood ejected by the heart per minute. It is the single most important number in cardiovascular physiology because it represents the rate at which the heart delivers oxygen and nutrients to every tissue in the body.

$CO = HR \times SV$ is an identity, as established in Chapter 1. But this identity, while correct, tells us nothing about how to measure cardiac output — and measurement is where physiology becomes clinical medicine.

The challenge is fundamental: you cannot directly measure flow in the aorta without an invasive catheter and flow probe. Everything done clinically to assess cardiac output is an inference — a calculation based on a conservation principle. Three methods dominate clinical practice, each based on a different form of the same underlying law:

- ***Fick principle*** *(oxygen consumption): the oldest and most physiologically rigorous*

- ***Thermodilution*** *(indicator dilution): the clinical standard for three decades*

- ***Pulse contour analysis*** *(waveform mathematics): the modern non-invasive approach*

All three are conservation laws in disguise. This chapter derives each from first principles.

Before the equations: why oxygen delivery is the purpose of the cardiovascular system.

Every organ has a different oxygen demand, a different oxygen extraction fraction, and a different tolerance for oxygen deficit. Understanding the Fick equation requires knowing what the tissues are actually doing with the oxygen they receive.

Haemoglobin: the molecule that makes it possible. *Blood carries 70-fold more oxygen per unit volume than water because of haemoglobin — a tetrameric protein (two α-chains encoded by HBA1/HBA2, two β-chains encoded by HBB) with four haem groups, each capable of binding one O_2 molecule. The cooperative binding of O_2 (Hill coefficient $n \approx 2.8$, from the $T{\to}R$ quaternary structure transition) produces the sigmoid oxygen-haemoglobin dissociation curve: haemoglobin is nearly fully saturated at normal arterial*

P_{O_2} *(100 mmHg, $SaO_2 \approx 98\%$) and releases substantial oxygen when venous P_{O_2} falls to 40 mmHg ($SvO_2 \approx 75\%$). Without cooperativity, haemoglobin would be either always saturated or always empty — useless for controlled oxygen delivery.*

The mitochondrion: why cells need the oxygen. *Oxygen is the terminal electron acceptor of the mitochondrial electron transport chain. Complex IV (cytochrome c oxidase) transfers electrons from cytochrome c to O_2, reducing it to water. This drives the proton gradient across the inner mitochondrial membrane that ATP synthase (Complex V) uses to produce ATP. Without oxygen, this chain stops. Cells fall back on anaerobic glycolysis (2 ATP per glucose) and begin accumulating lactate.*

The Fick equation as a conservation law. *The Fick equation does not assume anything about the heart or the circulation. It states only that oxygen cannot be created or destroyed: whatever oxygen the blood picks up in the lungs must equal what the tissues consume. The simplicity of this argument is its strength — it works regardless of valvular anatomy, shunt fractions, or cardiac rhythm.*

6.2   The Observation: Adolf Fick's Insight

In 1870, Adolf Fick proposed a deceptively simple idea: if you know how much oxygen the body consumes per minute, and you know how much more oxygen the arterial blood carries than the venous blood returning to the heart, then you can calculate exactly how much blood the heart must be pumping per minute to account for the difference.

The logic is a conservation of mass argument. Oxygen entering the systemic circulation per minute must equal oxygen consumed by the tissues. If blood arrives carrying C_a mL O_2/dL and leaves carrying C_v mL O_2/dL, then each decilitre of blood delivers $C_a - C_v$ mL of oxygen. If the tissues consume $\dot{V}O_2$ mL O_2/min, then:

$$Flow = \frac{O_2 \ consumed \ per \ minute}{O_2 \ delivered \ per \ dL}$$

This is the Fick principle, and it is one of the most important equations in clinical medicine.

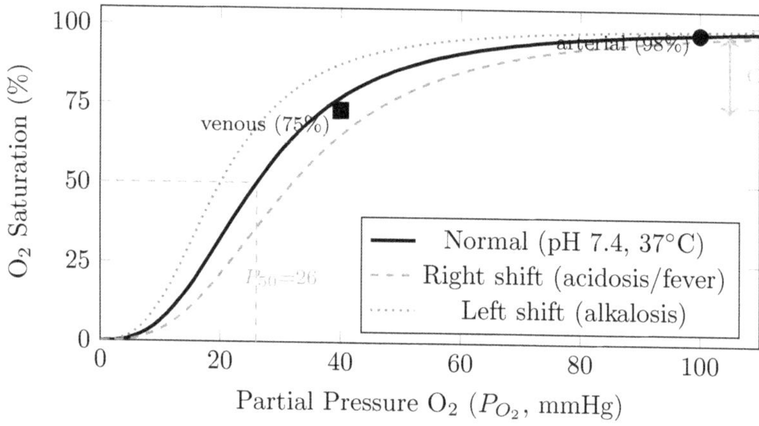

Figure 6.1: Oxygen-haemoglobin dissociation curve. The sigmoid shape arises from cooperative O_2 binding (Hill $n \approx 2.8$): haemoglobin is nearly fully saturated at arterial P_{O_2} (100 mmHg) and releases $\approx 25\%$ of its oxygen at resting venous P_{O_2} (40 mmHg). Right shift (acidosis, fever, high CO_2, 2,3-DPG) facilitates O_2 release to active tissues. The $C_{a-v}O_2$ difference (orange bracket) is the denominator of the Fick equation.

6.3 The Mathematics: Three Methods, One Law

6.3.1 Method 1 — The Fick Principle

Let $\dot{V}O_2$ (mL O_2/min) be whole-body oxygen consumption, C_a (mL O_2/dL) be arterial oxygen content, and C_v (mL O_2/dL) be mixed venous oxygen content (sampled from the pulmonary artery, where blood from all veins has mixed). The Fick equation is:

[Core Principle]

$$CO = \frac{\dot{V}O_2}{C_a - C_v} \times 10 \qquad (6.1)$$

The factor of 10 converts dL to L (since $C_a - C_v$ is in mL/dL and CO is *in L/min).*

Oxygen content C_a and C_v are computed from haemoglobin concentration and oxygen saturation:

[Useful Approximation]

$$C = 1.34 \cdot [Hb] \cdot SaO_2 + 0.003 \cdot P_{O_2} \tag{6.2}$$

where 1.34 mL O_2/g is the oxygen-carrying capacity of haemoglobin (Hüfner constant), [Hb] is haemoglobin concentration in g/dL, SaO_2 is fractional oxygen saturation, and the dissolved oxygen term $(0.003 \cdot P_{O_2})$ is small and often neglected.

Typical values at rest:

- $\dot{V}O_2 = 250$ *mL O_2/min*

- $C_a = 20$ *mL O_2/dL ($SaO_2 = 98\%$, Hb = 15 g/dL)*

- $C_v = 15$ *mL O_2/dL ($SvO_2 = 75\%$)*

- $CO = 250/(20 - 15) \times 10^{-1} = 250/5 = 50$ *dL/min = 5 L/min*

6.3.2   Method 2 — Thermodilution

Thermodilution uses cold saline as a thermal indicator. A bolus of cold saline (volume V_i, temperature T_i) is injected into the right atrium. A thermistor in the pulmonary artery records the temperature-time curve $T(t)$ as the cold mixes with warm blood.

By conservation of heat, the heat added by the cold bolus equals the heat removed from the blood:

$$CO = \frac{V_i \cdot (T_b - T_i) \cdot \rho_i c_i}{\rho_b c_b \int_0^\infty \Delta T(t) \, dt} \tag{6.3}$$

where T_b is blood temperature, $\rho_i c_i$ is the specific heat capacity of the injectate, $\rho_b c_b$ is the specific heat capacity of blood, and $\int_0^\infty \Delta T(t) \, dt$ is the area under the temperature-time curve.

The Stewart-Hamilton equation, equation (6.3), is an indicator dilution principle: higher CO washes the cold indicator through more quickly, producing a smaller area under the curve. Lower CO produces a larger, broader curve. The integral of temperature change is inversely proportional to cardiac output.

6.3.3   Method 3 — Pulse Contour Analysis

Modern pulse contour analysis methods (PiCCO, LiDCO, FloTrac) estimate CO from the arterial pressure waveform without a pulmonary artery catheter. The principle: the area under the systolic portion of the arterial pressure waveform is proportional to stroke volume, which multiplied by heart rate gives CO.

$$SV \propto \frac{1}{TPR} \int_{t_d}^{t_s} P(t) \, \mathrm{d}t \tag{6.4}$$

where the integral is taken from diastolic pressure t_d to the dicrotic notch t_s (end of systole), and TPR is estimated from the Windkessel model parameters (Chapter 3). The proportionality constant is calibrated by a reference CO measurement (thermodilution or Fick) at the start of monitoring, then updated continuously.

6.4   The Derivation: The Fick Equation as a Conservation Law

The Fick principle is a statement of conservation of mass applied to oxygen. Consider the lung as a steady-state system with the following flows:

- *Deoxygenated blood enters the pulmonary circulation at rate CO, carrying C_v mL O_2/dL*

- *Oxygenated blood leaves at rate CO, carrying C_a mL O_2/dL*

- *Oxygen enters from inspired air at rate $\dot{V}O_2$ mL O_2/min*

At steady state, oxygen in = oxygen out:

$$\dot{V}O_2 + CO \cdot C_v \; = \; CO \cdot C_a \tag{6.5}$$

Rearranging:

$$\dot{V}O_2 \; = \; CO \cdot (C_a - C_v) \tag{6.6}$$

$$CO \; = \; \frac{\dot{V}O_2}{C_a - C_v} \tag{6.7}$$

*This is conservation of mass. No assumptions about pump mechanics,
valve function, or heart rate are required. The Fick principle works for
any cardiac anatomy — including congenital heart disease with intracardiac shunts, where it can even quantify the shunt fraction.*

6.4.1   The Reverse Fick Method

*In the intensive care unit, directly measuring $\dot{V}O_2$ (requires a metabolic
cart) is often impractical. The* reverse Fick *method estimates $\dot{V}O_2$ from
CO, C_a, and C_v:*

$$\dot{V}O_2 \; = \; CO \cdot (C_a - C_v) \tag{6.8}$$

*When CO is measured by thermodilution, C_a from arterial blood gas, and
C_v from central venous blood gas, the reverse Fick gives a continuous
estimate of oxygen consumption — useful for monitoring tissue oxygen
demand in critical illness.*

6.5   The Clinical Interpretation: What CO Tells You and What It Does Not

6.5.1   Cardiac Index: Normalising for Body Size

*Raw CO (L/min) must be normalised for body surface area (BSA) to
compare across patients:*

$$CI = \frac{CO}{BSA} \quad [L/min/m^2] \tag{6.9}$$

BSA is estimated by the Du Bois formula:

$$BSA = 0.007184 \cdot W^{0.425} \cdot H^{0.725} \tag{6.10}$$

where W is weight in kg and H is height in cm. Normal CI: 2.5–4.0 $L/min/m^2$. CI < 2.2 indicates cardiogenic shock. CI < 1.8 is consistent with haemodynamic compromise requiring immediate intervention.

6.5.2 Mixed Venous Saturation: The Body's Report Card

Mixed venous oxygen saturation (SvO_2, measured in the pulmonary artery) tells you whether the body's oxygen supply matches its demand. From the Fick equation rearranged:

$$SvO_2 = SaO_2 - \frac{\dot{V}O_2}{1.34 \cdot [Hb] \cdot CO \cdot 10} \tag{6.11}$$

Normal SvO_2 is 65–75%. SvO_2 < 60% indicates that the tissues are extracting more oxygen than normal — either because CO has fallen, haemoglobin has fallen, or oxygen demand has risen.

SvO_2 encodes information about all four determinants of oxygen delivery simultaneously: CO, haemoglobin, saturation, and consumption. It is the single most informative number in haemodynamic monitoring.

__The four causes of low SvO_2:__ From equation (6.11), SvO_2 falls when: (1) CO falls (pump failure, hypovolaemia); (2) [Hb] falls (anaemia, haemorrhage); (3) SaO_2 falls (respiratory failure); (4) $\dot{V}O_2$ rises (fever, agitation, sepsis, exercise). Identifying which of the four is dominant determines the treatment. This is why the Fick equation, a 150-year-old conservation law, remains the foundation of modern haemodynamic management.

6.5.3 Oxygen Delivery and Extraction

Define oxygen delivery DO_2 as the rate at which the heart delivers oxygen to the tissues:

$$DO_2 \;=\; CO \cdot C_a \cdot 10 \quad [mL \ O_2/min] \tag{6.12}$$

Normal $DO_2 \approx 1{,}000 \ mL \ O_2/min$. Oxygen extraction ratio (OER):

$$OER \;=\; \frac{\dot{V}O_2}{DO_2} \;=\; \frac{C_a - C_v}{C_a} \tag{6.13}$$

Normal $OER \approx 25\%$ — the tissues extract one quarter of delivered oxygen. In shock, OER rises as the tissues extract more to compensate for reduced delivery. When OER exceeds 50–60%, the tissues are at their extraction limit and lactic acidosis develops — the biochemical marker of oxygen debt.

6.6 Factors That Determine Cardiac Output

$CO = HR \times SV$ is the identity. The physiology provides the control relationships:

6.6.1 Heart Rate

HR is set by the intrinsic pacemaker rate of the sinoatrial node (≈ 100 bpm) modulated by the autonomic nervous system. Sympathetic activation increases HR (positive chronotropy); parasympathetic activation decreases it.

The relationship between HR and CO is not linear. At very high HR (> 160–180 bpm), diastolic filling time is so short that EDV falls and SV drops, partially offsetting the HR increase. Maximum CO occurs at an optimal HR that depends on the individual's stroke volume and diastolic compliance.

6.6.2 Stroke Volume Determinants

From Chapters 4 and 5: $SV = EDV - V_d - P_{AO}/E_{es}$

- **Preload** *(EDV): determined by venous return, blood volume, and posture. Increased by fluid loading, leg elevation, recumbent posture.*

- **Afterload** *(P_{AO}): determined by MAP and TPR. Increased by hypertension, aortic stenosis, vasoconstrictors. Decreased by vasodilators, sepsis.*

- **Contractility** *(E_{es}): increased by sympathetic activation, inotropes (dobutamine, adrenaline). Decreased by myocardial ischaemia, heart failure, negative inotropes (β-blockers, calcium channel blockers).*

6.7 The Worked Example: Diagnosing Shock with the Fick Equation

A 72-year-old woman with known ischaemic cardiomyopathy (EF 25%) is admitted with worsening dyspnoea and hypotension. Pulmonary artery catheter measurements:

Parameter	Value	Normal
Arterial SaO_2	*97%*	*95–99%*
Mixed venous SvO_2	*42%*	*65–75%*
Haemoglobin	*11.2 g/dL*	*12–16 g/dL*
PA wedge pressure	*28 mmHg*	*<18 mmHg*

$\dot{V}O_2$ *estimated from body surface area at rest:* $BSA = 0.007184 \times 68^{0.425} \times 162^{0.725} \approx 1.74\ m^2$, $\dot{V}O_2 \approx 125 \times 1.74 = 218\ mL\ O_2/min$.

Step 1 — Oxygen content.

$$C_a = 1.34 \times 11.2 \times 0.97 = 14.6\ mL\ O_2/dL$$
$$C_v = 1.34 \times 11.2 \times 0.42 = 6.3\ mL\ O_2/dL$$

Step 2 — Cardiac output by Fick.

$$CO = \frac{218}{(14.6 - 6.3) \times 10} = \frac{218}{83} \approx 2.6 \; L/min$$

Step 3 — Cardiac index.

$$CI = \frac{2.6}{1.74} = 1.49 \; L/min/m^2 \quad (normal \geq 2.5)$$

CI of 1.49 confirms severe cardiogenic shock.

Step 4 — Oxygen delivery and extraction.

$$DO_2 = 2.6 \times 14.6 \times 10 = 380 \; mL \; O_2/min \quad (normal \approx 1{,}000)$$

$$OER = \frac{C_a - C_v}{C_a} = \frac{14.6 - 6.3}{14.6} = 57\% \quad (critical \; threshold)$$

OER of 57% exceeds the extraction limit. This patient is in oxygen debt and at high risk of lactic acidosis. Immediate intervention is needed: inotropic support, consideration of mechanical circulatory support.

Step 5 — Treatment targets. *Target $CI \geq 2.2 \; L/min/m^2$, $SvO_2 \geq 65\%$. Required CO increase: $2.2 \times 1.74 = 3.83 \; L/min$. Dobutamine target: increase CO from 2.6 to $\geq 3.8 \; L/min$ — a 46% increase in CO, achievable with moderate dobutamine doses (5–10 $\mu g/kg/min$).*

Model Assumptions, Chapter 6. *What the Fick equation assumes: (1) Steady-state oxygen consumption (no acute change in metabolic rate during measurement). (2) Complete mixing of venous blood at the pulmonary artery sampling site. (3) No intracardiac shunt. Where it breaks down: Intracardiac shunts (ASD, VSD) require a modified Fick equation with separate pulmonary and systemic flow terms. Thermodilution overestimates CO in tricuspid regurgitation (indicator recirculates). Pulse contour methods require calibration and drift with changes in vascular tone. All three methods have $\pm$10–15% measurement uncertainty.*

6.8   The Software Module

Module 6.1 — Cardiac Output Calculator and Oxygen Delivery Analyser

Input: *Method choice (Fick / thermodilution / pulse contour); arterial SaO_2; mixed venous SvO_2 (or $ScvO_2$ as surrogate); haemoglobin (g/dL); weight (kg); height (cm); $\dot{V}O_2$ if measured; thermodilution curve data if available.*

Processing:

1. *Compute C_a and C_v from haemoglobin and saturations (equation 6.2).*

2. *If $\dot{V}O_2$ known: compute CO by Fick (equation 6.1).*

3. *If thermodilution curve provided: integrate area under curve; compute CO by Stewart-Hamilton (equation 6.3).*

4. *Compute BSA, CI, DO_2, OER, SvO_2 (equations 6.9–6.13).*

5. *Classify haemodynamic state: normal, hyperdynamic, cardiogenic shock, distributive shock.*

6. *Compute treatment targets (CI $\geq$ 2.2, $SvO_2 \geq 65\%$) and required CO change.*

Output: *CO, CI, DO_2, OER, SvO_2; haemodynamic classification; treatment targets; sensitivity analysis (what if Hb rises 1 g/dL? what if SaO_2 improves to 99%?).*

Key equations:

$$CO = \dot{V}O_2/(C_a - C_v)$$

$$DO_2 = CO \cdot C_a \cdot 10$$

$$OER = (C_a - C_v)/C_a$$

Full   implementation   at   themathematicsoftheliving-

body.com/modules.

6.9 Chapter Summary

1. *Cardiac output cannot be measured directly — it is always inferred from a conservation law. The Fick principle (mass conservation), thermodilution (heat conservation), and pulse contour analysis (momentum conservation) are three manifestations of this single principle.*

2. *The Fick equation $CO = \dot{V}O_2/(C_a - C_v)$ is derived from oxygen mass balance at the lung. It requires no assumptions about cardiac mechanics and works for any anatomy including congenital heart disease.*

3. *Oxygen content $C = 1.34 \cdot [Hb] \cdot SaO_2$ links the Fick equation to bedside measurements. The four causes of low SvO_2 — low CO, low Hb, low SaO_2, high $\dot{V}O_2$ — follow directly from equation (6.11).*

4. *Cardiac index $CI = CO/BSA$ normalises for body size. Normal is 2.5–4.0 $L/min/m^2$. $CI < 1.8$ indicates haemodynamic compromise.*

5. *$OER = (C_a - C_v)/C_a$ measures the fraction of delivered oxygen extracted. $OER > 50\%$ signals the tissue extraction limit and impending oxygen debt.*

6. *In the worked example, Fick gave $CI = 1.49$ and $OER = 57\%$ — both indicating critical cardiogenic shock with oxygen debt, requiring a 46% increase in CO to reach safe targets.*

Key Equations, Chapter 6

$$CO = \frac{\dot{V}O_2}{C_a - C_v} \quad \text{(Fick principle)}$$

$$C = 1.34 \cdot [Hb] \cdot SaO_2 \quad \text{(oxygen content)}$$

$$DO_2 = CO \cdot C_a \cdot 10 \quad \text{(oxygen delivery)}$$

$$OER = \frac{C_a - C_v}{C_a} \quad \text{(extraction ratio)}$$

$$CI = \frac{CO}{BSA} \quad \text{(cardiac index)}$$

Key References, Chapter 6

1. *Fick, A. (1870). Über die Messung des Blutquantums in den Herzventrikeln. Sitz. d. Physik.-Med. Ges. zu Würzburg. [Original Fick principle statement.]*

2. *Wessel, H.U. et al. (1971). Limitations of thermal dilution curves for cardiac output determinations. J. Appl. Physiol. 30, 643–652. [Thermodilution accuracy and limitations.]*

3. *Rivers, E. et al. (2001). Early goal-directed therapy in the treatment of severe sepsis and septic shock. N. Engl. J. Med. 345, 1368–1377. [ScvO_2 as a treatment target in sepsis.]*

Chapter 7

The Action Potential: Hodgkin-Huxley in the Myocyte

7.1 The Physiology: Electricity Before Mechanics

Everything in Chapters 4, 5, and 6 — the pressure-volume loop, the Frank-Starling curve, cardiac output — depends on one thing happening first: the myocyte must be electrically excited. Without the action potential, no calcium is released, no cross-bridges form, no force is generated.

The cardiac action potential is the electrical event that initiates every heartbeat. It is a precisely timed sequence of ion movements across the myocyte membrane, each carried by a specific channel protein, each mathematically describable. Understanding it at this level explains:

- *Why the QT interval on the ECG predicts arrhythmia risk — and precisely which channel is responsible*

- *Why digoxin slows the heart and increases contractility through the same mechanism*

- *Why hyperkalaemia causes the tall peaked T-waves seen on ECG*

- *Why calcium channel blockers and beta-blockers affect the heart differently despite both slowing the rate*

The mathematical framework is the Hodgkin-Huxley model, originally derived for the squid giant axon. Adapted for the cardiac myocyte, it is the foundation of computational cardiology and the basis of every drug-channel interaction prediction in pharmaceutical development.

Before the equations: the protein architecture of the ion channel.

Ion channels are not mere pores. They are allosteric molecular machines with precisely engineered gating mechanisms. Understanding their protein structure makes the Hodgkin-Huxley gating variables physically interpretable rather than abstract mathematical constructs.

The voltage-gated sodium channel (Nav1.5, encoded by SCN5A). *Nav1.5 is a single α-subunit protein with four homologous domains (DI–DIV), each containing six transmembrane helices (S1–S6). The S4 helix in each domain contains 4–7 positively charged arginine and lysine residues. When the membrane depolarises, the electric field drives these charged residues outward — a physical rotation and translation of the S4 helix that is the molecular basis of* *the activation gate m. The pore is formed by the S5-S6 loops of all four domains, with the selectivity filter (DEKA motif: Asp-Glu-Lys-Ala) conferring Na^+ selectivity over K^+. Inactivation (the h gate) is mediated by the cytoplasmic III-IV linker, which acts as a hinged lid that occludes the pore from the intracellular side within milliseconds of activation.*

The hERG channel (I_{Kr}, encoded by KCNH2). *The rapid delayed rectifier K^+ channel is unusual: it activates and inactivates rapidly, but inactivation is relieved on repolarisation faster than deactivation. This kinetic quirk — slow deactivation — is what makes hERG the dominant repolarising current during Phase 3. The channel's deep, hydrophobic drug-binding pocket (formed by S6 helices of all four subunits) makes it exceptionally susceptible to block by structurally diverse drugs, explaining why so many non-cardiac drugs prolong the QT interval.*

Why gating variables work: *The Hodgkin-Huxley gating variable m is the fraction of S4 helices in the activated conformation. The h gate is the fraction of III-IV linkers not occluding the pore. The product $m^3 h$ in equation (7.6) reflects the cooperativity of three independent S4 movements required for channel opening, times the probability the inactivation gate is open.*

7.2 The Observation: Five Phases of the Ventricular Action Potential

The ventricular myocyte action potential has five distinct phases, each dominated by specific ion currents:

Phase 0 — Rapid depolarisation. *A stimulus brings the membrane potential above the threshold (≈ -65 mV). Fast sodium channels (I_{Na}) open explosively. Na^+ rushes in. Membrane potential rises from -85 mV to $+20$ mV in < 2 ms. This is the fastest electrical event in the body.*

Phase 1 — Early repolarisation. *Fast Na^+ channels inactivate rapidly. Transient outward K^+ current (I_{to}) briefly repolarises the mem-*

brane to ≈ 0 mV, producing the characteristic notch on the action potential waveform.

Phase 2 — Plateau. *The defining feature of the cardiac action potential. L-type calcium channels ($I_{Ca,L}$) open, bringing Ca^{2+} into the cell. This inward current balances the outward K^+ current (I_{Kr}, I_{Ks}), holding the membrane potential near 0 mV for 200–300 ms. The plateau is what triggers calcium-induced calcium release from the sarcoplasmic reticulum — the link between electricity and contraction.*

Phase 3 — Rapid repolarisation. *L-type calcium channels inactivate. Delayed rectifier K^+ currents (I_{Kr}, I_{Ks}) fully activate and repolarise the membrane back to -85 mV.*

Phase 4 — Resting membrane potential. *The membrane potential is stable at ≈ -85 mV, maintained by the inward rectifier K^+ current (I_{K1}). The Na^+/K^+ ATPase slowly restores ionic gradients.*

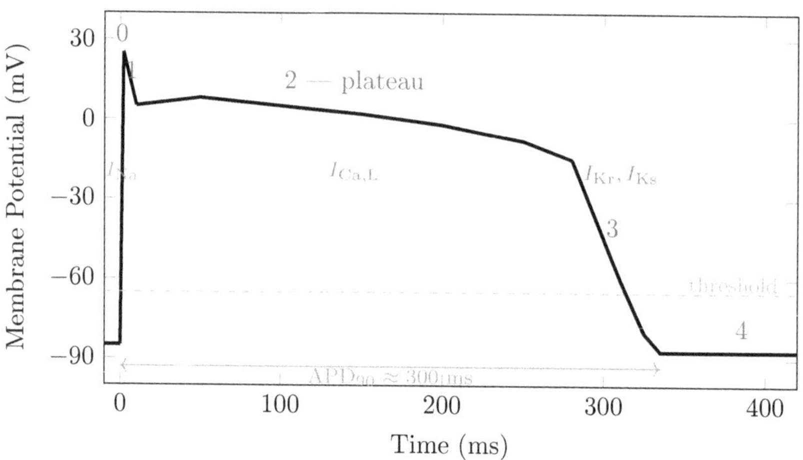

Figure 7.1: The ventricular myocyte action potential. Phase 0: rapid depolarisation (I_{Na}, fast Na^+ channels, <2 ms). Phase 1: early repolarisation (I_{to}). Phase 2: plateau ($I_{Ca,L}$ balances I_{Kr}, 200–250 ms). Phase 3: rapid repolarisation (I_{Kr}, I_{Ks}). Phase 4: resting potential (-85 mV, I_{K1}). APD$_{90}$ corresponds to the QT interval.

7.3 The Mathematics: The Hodgkin-Huxley Framework

7.3.1 The Membrane as a Circuit

Model the myocyte membrane as an electrical circuit: a capacitor (the lipid bilayer, capacitance $C_m \approx 1$ $\mu F/cm^2$) in parallel with conductance elements (the ion channels). The membrane current equation is:

$$C_m \frac{dV}{dt} = -I_{ion}(V,t) + I_{stim}(t) \tag{7.1}$$

where V is the membrane potential (mV), I_{ion} is the total ionic current (sum of all channel currents), and I_{stim} is an applied stimulus current. Each ionic current follows Ohm's law:

$$I_j = g_j(V,t) \cdot (V - E_j) \tag{7.2}$$

where g_j is the channel conductance and E_j is the reversal potential (Nernst potential) for ion j.

7.3.2 The Nernst Equation: Reversal Potentials

The reversal potential for each ion is set by its concentration gradient across the membrane, given by the Nernst equation:

[Core Principle]

$$E_j = \frac{RT}{z_j F} \ln\left(\frac{[ion]_o}{[ion]_i}\right) \tag{7.3}$$

where R is the gas constant, T is temperature, z_j is the ionic charge, F is the Faraday constant, and the subscripts o and i denote extracellular and intracellular concentrations. At body temperature (37° C), $RT/F = 26.7$ mV.

Ion	$[\cdot]_o$ (mM)	$[\cdot]_i$ (mM)	z	E_j (mV)
Na^+	145	12	+1	+67
K^+	4	150	+1	−94
Ca^{2+}	2	10^{-4}	+2	+133
Cl^-	120	30	−1	−37

Na^+ and Ca^{2+} have positive reversal potentials — inward currents that depolarise the membrane. K^+ has a negative reversal potential — outward current that repolarises it. This is why Na^+/Ca^{2+} entry drives depolarisation and K^+ exit drives repolarisation.

7.3.3 Gating Variables: Hodgkin-Huxley Gates

Channel conductance is controlled by gating variables — dimensionless quantities between 0 (channel closed) and 1 (channel fully open). Each gate satisfies a first-order ODE:

[Useful Approximation]

$$\frac{\mathrm{d}x}{\mathrm{d}t} = \alpha_x(V)(1-x) - \beta_x(V)x \tag{7.4}$$

where $\alpha_x(V)$ is the voltage-dependent opening rate and $\beta_x(V)$ is the closing rate. The steady-state value and time constant are:

$$x_\infty(V) = \frac{\alpha_x}{\alpha_x + \beta_x}, \qquad \tau_x(V) = \frac{1}{\alpha_x + \beta_x} \tag{7.5}$$

7.3.4 The Fast Sodium Current I_{Na}

The fast sodium channel has three gates: activation gate m (opens fast), and two inactivation gates h (fast) and j (slow):

$$I_{Na} = \bar{g}_{Na} \cdot m^3 \cdot h \cdot j \cdot (V - E_{Na}) \tag{7.6}$$

The cubic power of m reflects the cooperative opening of three independent *activation subunits. The h gate inactivates the channel within milliseconds of opening — this is why Phase 0 is so brief and why there is an absolute refractory period during which the channel cannot reopen.*

Rate functions at body temperature (from Luo-Rudy formulation, the standard cardiac Hodgkin-Huxley model):

$$\alpha_m = \frac{0.32(V + 47.13)}{1 - e^{-0.1(V+47.13)}} \tag{7.7}$$

$$\beta_m = 0.08 \, e^{-V/11} \tag{7.8}$$

$$\alpha_h = 0.135 \, e^{-(V+80)/6.8} \tag{7.9}$$

$$\beta_h = \frac{3.56 \, e^{0.079V} + 3.1 \times 10^5 e^{0.35V}}{1} \tag{7.10}$$

7.3.5 The L-type Calcium Current $I_{\text{Ca,L}}$

The L-type calcium channel is the key link between electrical activity and mechanical contraction. It carries inward Ca^{2+} current during the plateau and triggers calcium-induced calcium release (CICR) from the sarcoplasmic reticulum:

$$I_{Ca,L} = \bar{g}_{Ca} \cdot d \cdot f \cdot f_{Ca} \cdot (V - E_{Ca}) \tag{7.11}$$

where d is the activation gate, f is the voltage-dependent inactivation gate, and f_{Ca} is the calcium-dependent inactivation gate (the channel inactivates when intracellular Ca^{2+} rises, providing a negative feedback mechanism).

7.3.6 The Delayed Rectifier K$^+$ Currents

Two delayed rectifier potassium currents repolarise the ventricle:

$$I_{Kr} = \bar{g}_{Kr} \cdot x_r \cdot r \cdot (V - E_K) \tag{7.12}$$

$$I_{Ks} = \bar{g}_{Ks} \cdot x_s^2 \cdot (V - E_K) \tag{7.13}$$

I_{Kr} *(rapidly activating) is encoded by the hERG gene. This is the channel most commonly blocked by drugs causing QT prolongation — including many antibiotics, antipsychotics, and antihistamines. Blocking I_{Kr} prolongs Phase 3 repolarisation, lengthening the QT interval and risking torsades de pointes.*

7.4   The Derivation: QT Interval from Channel Kinetics

The QT interval on the ECG (from Q wave to end of T wave) represents ventricular depolarisation and repolarisation. Its duration corresponds to the action potential duration (APD) of the ventricular myocyte. From the Hodgkin-Huxley model, APD is determined by the time for the membrane potential to return from peak to -65 mV (the threshold):

$$APD_{90} = \int_0^{t_{90}} dt \quad s.t. \quad V(t_{90}) = -65 \ mV \tag{7.14}$$

The APD is set by the balance of inward (depolarising) and outward (repolarising) currents during the plateau. Specifically, APD is prolonged when:

1. *Outward K^+ currents are reduced (I_{Kr} block by drugs, hypokalaemia reducing I_{K1})*

2. *Inward currents are increased (late Na^+ current in LQTS type 3, increased Ca^{2+} in hypercalcaemia)*

3. *Heart rate is slow (at slow rates, K^+ channels have more time to close between beats, reducing available repolarisation reserve — Bazett's correction)*

Bazett's formula *corrects the QT interval for heart rate:*

[Approximation — see Model Assumptions]

$$QTc = \frac{QT}{\sqrt{RR}} \tag{7.15}$$

where RR is the R-R interval in seconds. $QTc > 440$ ms in men and > 460 ms in women is considered prolonged by convention, though the normal range has substantial overlap with the abnormal range and no single threshold reliably predicts arrhythmia risk in an individual. $QTc > 500$ ms carries significant risk of torsades de pointes.

Bazett's formula is an empirical fit. Its mathematical justification comes from the Hodgkin-Huxley model: at longer cycle lengths (slower heart rates), the gating variables of I_{Kr} have time to recover more completely, but the steady-state APD also lengthens. The $\sqrt{RR}$ dependence approximates the nonlinear relationship between cycle length and APD that emerges from the gate dynamics.

7.5 The Clinical Interpretation: Drugs, Channels, and the ECG

7.5.1 Drug-Channel Interactions

Every antiarrhythmic drug works by modifying one or more Hodgkin-Huxley gating variables:

Drug class	Target	Effect	ECG
Class Ia	I_{Na}	*Reduces $\bar{g}_{Na}$, slows Phase 0*	$\uparrow QRS$, $\uparrow QT$
Class Ib	I_{Na}	*Shortens APD in ischaemic tissue*	$\downarrow QT$
Class Ic	I_{Na}	*Marked Phase 0 slowing*	$\uparrow\uparrow QRS$
Class II	*β-receptor*	*Reduces $I_{Ca,L}$, slows SAN*	$\downarrow HR$
Class III	I_{Kr}	*Prolongs Phase 3*	$\uparrow QT$
Class IV	$I_{Ca,L}$	*Reduces plateau, slows SAN*	$\downarrow HR$, $PR\uparrow$

7.5.2 Electrolyte Effects on Channel Kinetics

Electrolyte abnormalities alter channel kinetics in precisely predictable ways:

Hyperkalaemia. *High extracellular K^+ reduces E_K (less negative by Nernst equation), reducing the driving force for outward K^+ current. Paradoxically, mild hyperkalaemia (5–6 mEq/L) accelerates Phase 3 repolarisation (peaked T-waves); severe hyperkalaemia (> 7 mEq/L) depolarises the resting membrane, inactivating I_{Na} channels and slowing conduction (wide QRS, sine-wave pattern).*

Hypokalaemia. *Low extracellular K^+ reduces I_{K1} (the inward rectifier stabilising Phase 4), making the resting potential less stable and the membrane more excitable. Also reduces I_{Kr} amplitude, prolonging APD and QT (U-waves, flat T-waves).*

Hypercalcaemia. *High Ca^{2+} shortens the plateau (enhanced f_{Ca} inactivation of $I_{Ca,L}$), shortening QT. Hypocalcaemia prolongs the plateau, lengthening QT.*

The QT interval is a window into channel kinetics: *Every drug and every electrolyte abnormality that affects the QT interval does so by altering one or more terms in the Hodgkin-Huxley ODE. $QTc > 500$ ms means the plateau is abnormally prolonged — Phase 3 repolarisation is failing to overcome the inward currents. This creates the substrate for early afterdepolarisations (EADs), which can trigger torsades de pointes. The mathematics specifies exactly which channel to target to correct it.*

7.6 The Sinoatrial Node: Automaticity as a Limit Cycle

The ventricular action potential described above requires an external stimulus. The sinoatrial node (SAN) generates its own rhythm — it fires spontaneously because it lacks the inward rectifier I_{K1} that stabilises Phase 4 in ventricular cells.

Instead, the SAN has the funny current I_f — a mixed Na^+/K^+ inward current that activates on hyperpolarisation (hence "funny" — it opens backwards compared to most channels). During Phase 4, I_f slowly depolarises the membrane toward threshold, generating the pacemaker potential:

$$C_m \frac{\mathrm{d}V}{\mathrm{d}t} = -I_f - I_{Ca,T} - I_{Ca,L} - I_{Kr} - I_{Ks} \qquad (7.16)$$

The SAN ODE has no stable equilibrium. Its solution is a limit cycle — a periodic orbit in phase space that the system returns to after any perturbation. The intrinsic cycle length (period of the limit cycle) sets the intrinsic heart rate (≈ 100 bpm when denervated).

Sympathetic activation increases I_f and $I_{Ca,L}$ conductances, steepening the pacemaker slope and shortening cycle length (increasing HR). Parasympathetic activation increases K^+ conductance (via I_{KAch}), hyperpolarising the membrane and flattening the pacemaker slope (decreasing HR).

7.7 The Worked Example: Predicting QTc from Drug Dose

A 61-year-old man with atrial fibrillation is started on amiodarone (a Class III antiarrhythmic, I_{Kr} blocker). Baseline $QT = 380$ ms, HR = 74 bpm (RR = 0.811 s). After 6 weeks at steady-state dose, $QT = 440$ ms, HR = 58 bpm (RR = 1.034 s).

Step 1 — Compute baseline QTc.

$$QTc_{baseline} = \frac{380}{\sqrt{0.811}} = \frac{380}{0.901} = 422 \ ms$$

Step 2 — Compute QTc on amiodarone.

$$QTc_{drug} = \frac{440}{\sqrt{1.034}} = \frac{440}{1.017} = 433 \ ms$$

Step 3 — Interpret. The raw QT lengthened by 60 ms, but most of this is rate correction — amiodarone slows the heart. The QTc increased by only 11 ms (422 to 433 ms), which is within acceptable limits (QTc < 500 ms, increase < 60 ms from baseline).

Step 4 — Channel interpretation. Amiodarone blocks I_{Kr} (Phase 3 delay) and $I_{Ca,L}$ (reduces plateau height). These effects partially cancel: I_{Kr} block prolongs APD; $I_{Ca,L}$ block shortens it. Net effect: moderate QT prolongation. This is why amiodarone, despite being a potent I_{Kr} blocker, causes less torsades de pointes than other Class III drugs — the calcium channel blockade provides partial protection.

Step 5 — Safety threshold. Continue monitoring. If QTc reaches 500 ms or increases by > 60 ms from baseline, reduce dose or discontinue.

Model Assumptions, Chapter 7. What the Hodgkin-Huxley model assumes: (1) The membrane is isopotential (space-clamped). (2) Ion channels are independent (no cooperativity between channels). (3) Ionic concentrations are constant (the model does not track Na^+/K^+ concentration changes). (4) Temperature is fixed at

37° C. Where it breaks down: Ion accumulation in the restricted cleft between myocytes (K^+ accumulation during tachycardia modifies gating). The Markov chain representation of real channel gating (hundreds of states) is better than the 2-gate Hodgkin-Huxley approximation. Bazett's formula is empirical and less accurate at extremes of heart rate.

7.8 The Software Module

Module 7.1 — Cardiac Action Potential Simulator (Luo-Rudy Model)

Input: *Model variant (ventricular / SAN); maximum conductances ($\bar{g}_{Na}$, $\bar{g}_{Ca}$, $\bar{g}_{Kr}$, $\bar{g}_{Ks}$); drug effect (fractional block of each conductance); extracellular ion concentrations (K^+, Na^+, Ca^{2+}); stimulation protocol (single beat / pacing train).*

Processing:

1. *Set up the Luo-Rudy ODE system (equation 7.1 with currents 7.6–7.13).*

2. *Compute Nernst potentials from ion concentrations (equation 7.3).*

3. *Integrate using scipy.integrate.solve_ivp (stiff solver LSODA); simulate 5 beats to reach steady state.*

4. *Extract APD_{90}, peak V, resting V, plateau duration, and upstroke velocity.*

5. *Compute QTc using Bazett's formula.*

6. *Apply drug block: reduce specified conductance by fraction; recompute APD and QTc.*

Output: *Action potential waveform; phase plot (dV/dt vs V); APD_{90}; QTc; each ionic current time course; drug-induced QTc change; safety flag if $QTc > 500$ ms.*

> *Full implementation at themathematicsoftheliving-body.com/modules.*

7.9 Chapter Summary

1. *The cardiac action potential is governed by the Hodgkin-Huxley ODE: $C_m \, dV/dt = -I_{ion} + I_{stim}$. Each ionic current obeys Ohm's law with a voltage- and time-dependent conductance.*

2. *Reversal potentials are set by the Nernst equation. Na^+ and Ca^{2+} drive depolarisation; K^+ drives repolarisation.*

3. *Gating variables (m, h, j, d, f, x_r, x_s) each satisfy a first-order ODE with voltage-dependent rates $\alpha(V)$ and $\beta(V)$. The QT interval reflects the time course of Phase 3 repolarisation, set by the balance of I_{Kr}, I_{Ks}, and $I_{Ca,L}$.*

4. *$QTc = QT/\sqrt{RR}$ corrects for heart rate. $QTc > 500$ ms or $\Delta QTc > 60$ ms from baseline signals dangerous APD prolongation and torsades de pointes risk.*

5. *The SAN generates automaticity because it lacks I_{K1} and has the funny current I_f, producing a pacemaker potential (slow Phase 4 depolarisation) whose mathematical structure is a limit cycle.*

6. *In the amiodarone worked example, raw QT lengthened 60 ms but QTc increased only 11 ms after rate correction — illustrating why Bazett's correction is essential for interpreting drug-induced QT changes.*

Key Equations, Chapter 7

$$C_m \frac{\mathrm{d}V}{\mathrm{d}t} = -I_{ion} + I_{stim} \quad \textit{(membrane ODE)}$$

$$E_j = \frac{RT}{z_j F} \ln\left(\frac{[ion]_o}{[ion]_i}\right) \quad \textit{(Nernst)}$$

$$\frac{\mathrm{d}x}{\mathrm{d}t} = \alpha_x(1-x) - \beta_x x \quad \textit{(gating variable)}$$

$$QTc = \frac{QT}{\sqrt{RR}} \quad \textit{(Bazett's formula)}$$

Key References, Chapter 7

1. *Hodgkin, A.L. & Huxley, A.F. (1952). A quantitative description of membrane current and its application to conduction and excitation in nerve. J. Physiol. 117, 500–544. [The foundational paper, Nobel Prize 1963.]*

2. *Luo, C.H. & Rudy, Y. (1991). A model of the ventricular cardiac action potential. Circ. Res. 68, 1501–1526. [The standard cardiac Hodgkin-Huxley formulation.]*

3. *Bazett, H.C. (1920). An analysis of the time-relations of electrocardiograms. Heart 7, 353–370. [Original derivation of $QTc = QT/\sqrt{RR}$.]*

Chapter 8

Arrhythmia as Bifurcation: When the Rhythm Breaks

8.1 The Physiology: Why Arrhythmias Are Not Random

An arrhythmia is not a random malfunction of the heart. It is a qualitative change in the behaviour of the cardiac dynamical system — a shift from one pattern of activity to another, triggered when a parameter crosses a critical threshold.

This is the mathematical concept of a bifurcation*: a sudden qualitative change in the long-term behaviour of a dynamical system as a parameter is varied continuously. The rhythm does not gradually become more irregular. It switches. A heart in sinus rhythm does not drift slowly into ventricular fibrillation. It is stable until a critical condition is met, then transitions abruptly.*

Understanding arrhythmias as bifurcations explains several observations that seem puzzling otherwise:

- *Why a brief ectopic beat can trigger sustained ventricular tachycardia in an ischaemic heart but not in a normal heart — the normal heart is not near the bifurcation point; the ischaemic heart is*

- *Why the same drug that terminates one arrhythmia can trigger another — it moves the system past a different bifurcation*

- *Why defibrillation works by delivering current to all cells simultaneously — resetting the system to a state from which the stable sinus rhythm attractor can recapture it*

- *Why some arrhythmias (re-entry) are self-sustaining and others (triggered activity) require repeated initiating events*

Before the equations: the genes behind the arrhythmia.

Arrhythmias are not abstract dynamical events. They arise from mutations, drug interactions, or electrolyte disturbances that alter specific ion channel proteins. The genetic basis of the inherited arrhythmia syndromes maps the bifurcation theory of Section 8.3 di-

rectly to molecular biology.

**Long QT syndrome (LQTS) — too much inward, too little
outward.** *LQTS type 1 (KCNQ1 loss-of-function): KCNQ1 encodes
the α-subunit of the slow delayed rectifier I_{Ks}. Loss of function reduces repolarisation reserve, prolonging APD (longer plateau, larger
excitation gap for EADs). LQTS type 2 (KCNH2/hERG loss-of-function): reduces I_{Kr}, the same mechanism drug-induced QT prolongation exploits. LQTS type 3 (SCN5A gain-of-function): late
Na^+ current ($I_{Na,late}$) persists into the plateau, increasing inward
current and prolonging APD. In all three types, the EAD condition
(equation 8.4) is met: inward exceeds outward during Phase 3.*

Brugada syndrome (SCN5A loss-of-function). *Reduced
Nav1.5 function shortens the action potential in the right ventricular epicardium (which has prominent I_{to}) but not in the endocardium
(which lacks I_{to}). This transmural voltage gradient during Phase 2
creates the substrate for phase-2 re-entry — a uniquely localised
form of re-entry that produces the Brugada ECG pattern (coved ST
elevation in V1–V2) and risk of polymorphic VT/VF.*

Catecholaminergic polymorphic VT (CPVT, RYR2 gain-of-function). *Ryanodine receptor 2 (RyR2) controls Ca^{2+} release
from the SR. Gain-of-function mutations cause diastolic SR Ca^{2+}
leak at high sympathetic tone (exercise or emotional stress), generating delayed afterdepolarisations (DADs) via NCX — exactly
the mechanism described in Section 8.2. CPVT is a pure DAD-mediated arrhythmia with a structurally normal heart and normal
resting ECG.*

The bifurcation theory tells us when *arrhythmia occurs; the molecular biology tells us* which protein *is responsible.*

8.2 The Observation: Three Mechanisms of Arrhythmia

All arrhythmias arise from one or more of three fundamental mechanisms:

*Mechanism 1 — **Abnormal automaticity.** A cell outside the SAN develops spontaneous phase 4 depolarisation — becomes an ectopic pacemaker. This can occur in Purkinje fibres, the AV node, or ventricular myocytes injured by ischaemia. The ectopic focus fires faster than the SAN and takes over the rhythm, or competes with it (parasystole).*

*Mechanism 2 — **Triggered activity.** Oscillations in membrane potential following a normal action potential — afterdepolarisations — reach threshold and trigger a new action potential. Two types:*

Early afterdepolarisations (EADs): Occur during the plateau or early Phase 3. Caused by reactivation of $I_{Ca,L}$ or late I_{Na} when APD is prolonged (long QT). The mechanism of torsades de pointes.

Delayed afterdepolarisations (DADs): Occur after full repolarisation. Caused by spontaneous Ca^{2+} release from an overloaded sarcoplasmic reticulum, activating the Na^+/Ca^{2+} exchanger (I_{NCX}), generating inward current. The mechanism of digitalis toxicity and catecholaminergic VT.

*Mechanism 3 — **Re-entry.** The same electrical impulse circulates repeatedly through a loop of tissue, re-exciting tissue that has recovered from the previous impulse. This is the most common mechanism of sustained tachyarrhythmias — atrial flutter, AVNRT, accessory pathway tachycardia, ventricular tachycardia.*

8.3 The Mathematics: Re-entry as a Topological Condition

8.3.1 The Re-entry Circuit

Re-entry requires three conditions, each with a precise mathematical statement:

*Condition 1 — **A circuit.** A closed path of excitable tissue with at least two limbs through which an impulse can travel.*

*Condition 2 — **Unidirectional block.** One limb must be blocked in one direction but not the other. This is typically a region of slow conduction or inhomogeneous refractoriness.*

Condition 3 — Sufficient path length. *The circuit must be long enough that by the time the impulse completes one loop, the tissue at the starting point has recovered excitability.*

The critical condition is:

$$Path\ length > \lambda \ = \ CV \times ERP \tag{8.1}$$

where λ is the wavelength *of the cardiac impulse (the minimum path length for re-entry), CV is conduction velocity (m/s), and ERP is the effective refractory period (s).*

Normal values: $CV \approx 0.5$ m/s, $ERP \approx 0.25$ s, giving $\lambda \approx 0.125$ m $= 12.5$ cm. For re-entry to occur in the ventricle, the circuit must be at least 12.5 cm in circumference. A 10 cm scar from a prior MI does not support re-entry; a 15 cm scar does. This is the mathematical basis of the left ventricular substrate for VT.

8.3.2   Bifurcation from Sinus Rhythm to Re-entry

Define the excitation gap *as the difference between the circuit length L and the wavelength:*

$$Gap \ = \ L - \lambda \ = \ L - CV \times ERP \tag{8.2}$$

When $Gap \leq 0$: re-entry cannot be sustained. The impulse reaches refractory tissue before completing the circuit and terminates.

When $Gap > 0$: re-entry is self-sustaining. The impulse completes the circuit and finds recovered tissue.

The transition from $Gap \leq 0$ to $Gap > 0$ is a saddle-node *bifurcation — as the control parameter (gap) crosses zero, the system abruptly transitions from a stable fixed point (sinus rhythm) to a stable limit cycle (re-entrant tachycardia).*

Two ways a drug can reduce the excitation gap and terminate re-entry:

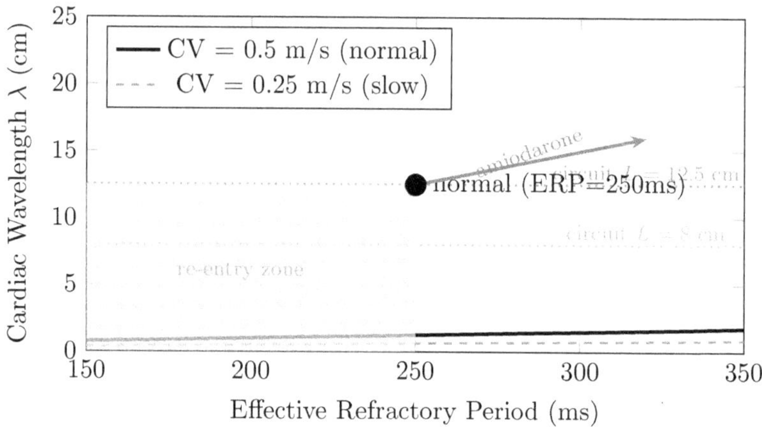

Figure 8.1: Cardiac wavelength $\lambda = \mathrm{CV} \times \mathrm{ERP}$ determines re-entry risk. If λ falls below the circuit length L (shaded zone), the excitation gap is positive and re-entry is self-sustaining. Normal parameters place the heart safely above the threshold ($\lambda \approx 12.5$ cm for a typical 12.5 cm circuit). Amiodarone increases ERP, raising λ out of the re-entry zone. Slow conduction (red dashed) reduces λ, enlarging the pro-arrhythmic zone.

1. **Reduce CV** *(Class I drugs, sodium channel blockers): reduces λ — paradoxically, slowing conduction* increases *the gap and may accelerate the tachycardia. This is why Class Ic drugs are proarrhythmic in structural heart disease.*

2. **Increase ERP** *(Class III drugs, potassium channel blockers): increases λ, eliminates the gap, terminates re-entry. This is the mechanism of amiodarone in VT.*

8.3.3   Atrial Fibrillation: Multiple Wavelets

Atrial fibrillation (AF) is not a single re-entrant circuit. It is multiple simultaneous wavelets of re-entry, each self-sustaining, constantly colliding and re-initiating. This is the multiple wavelet hypothesis *of Moe.*

The number of simultaneous wavelets N is approximately:

[Approximation — see Model Assumptions]

$$N \approx \frac{A_{atrium}}{\lambda^2} \tag{8.3}$$

where A_{atrium} is the atrial surface area and λ^2 approximates the area occupied by each wavelet. AF requires a critical number of simultaneous wavelets to be self-sustaining: if N falls below $\approx$ 3–4, AF terminates spontaneously (insufficient wavelets to prevent extinction by mutual collision).

This equation has three profound clinical implications:

Implication 1 — Why AF becomes persistent with time. Chronic AF causes atrial remodelling: the atrium enlarges (A_{atrium} increases) and the ERP shortens (λ decreases). Both changes increase N. "AF begets AF."

Implication 2 — Why cardioversion works. External electrical cardioversion simultaneously depolarises all atrial tissue, resetting all wavelets. When the tissue repolarises uniformly, N briefly falls to zero. If the underlying substrate has not been permanently altered, sinus rhythm resumes.

Implication 3 — Why ablation works. Pulmonary vein isolation reduces A_{atrium} (by electrically isolating a portion of the atrium), reducing N below the critical threshold. Linear ablation lesions further compartmentalise the atrium, reducing the effective area available for wavelet propagation.

8.4 The Derivation: Early Afterdepolarisations from the Hodgkin-Huxley Model

EADs arise from the Hodgkin-Huxley ODE when APD is sufficiently prolonged that $I_{Ca,L}$ can reactivate during Phase 3. The condition for EAD generation is a balance between inward and outward currents during the plateau:

At the bifurcation point, the net ionic current passes through zero during repolarisation:

$$I_{Ca,L}(V^*,t^*) + I_{Na,late}(V^*,t^*) \; = \; I_{Kr}(V^*,t^*) + I_{Ks}(V^*,t^*) \qquad (8.4)$$

At (V^*,t^*), the membrane potential is in unstable equilibrium during Phase 3. Any small perturbation that tips the balance toward net inward current produces an EAD — the membrane depolarises instead of continuing to repolarise.

This is a Hopf *bifurcation: as a parameter (APD, drug concentration, K^+ concentration) is varied, a stable fixed point (smooth repolarisation) loses stability and a limit cycle (EAD oscillations) is born. The oscillations can reach threshold and trigger a new action potential — the mechanism of torsades de pointes.*

Quantitatively, the risk of EAD generation increases when:

$$\frac{I_{Ca,L} + I_{Na,late}}{I_{Kr} + I_{Ks}} \gtrsim 1 \qquad (8.5)$$

This ratio is directly computable from the Hodgkin-Huxley model with drug-modified conductances. It is the quantitative safety index for any drug that affects cardiac channels.

8.5 The Clinical Interpretation: Treating Arrhythmias with Bifurcation Theory

8.5.1 Terminating Re-entry: Shifting the Bifurcation Point

To terminate a re-entrant tachycardia, the excitation gap must be eliminated. Three clinical strategies map directly to the gap equation:

Strategy		*Parameter*	*Clinical tool*
Increase ERP		$\uparrow \lambda$	*Class III drugs, adenosine*
Reduce circuit	*cir-*	$\downarrow L$	*Ablation, surgery*
Reset cells	*all*	$Gap \to -\infty$	*Cardioversion, defibrillation*

8.5.2 Preventing EADs: Maintaining the Ratio

To prevent torsades de pointes, keep the inward/ outward current ratio below 1. Four strategies:

1. **Avoid QT-prolonging drugs** *(reduces numerator: less late I_{Na}, less $I_{Ca,L}$ activation time)*

2. **Correct hypokalaemia** *(increases denominator: higher $[K^+]_o$ restores I_{Kr} amplitude)*

3. **Use mexiletine** *(Class Ib): blocks late I_{Na} specifically, reducing the numerator without prolonging QT further*

4. **Pace faster** *(shorter cycle length reduces APD, keeping ERP shorter than the EAD oscillation period)*

Why defibrillation works — the mathematical answer: *During ventricular fibrillation, the myocardium is partitioned into multiple independent wavefronts, each in a different phase of the action potential. A defibrillation shock delivers sufficient current (~ 200 J) to simultaneously raise the membrane potential of all cells above threshold — depolarising cells in all phases. When the shock ends, all cells are in their refractory period simultaneously. There are no recovered cells for re-entrant wavefronts to propagate into. The excitation gap is eliminated globally. If the underlying substrate is stable, sinus rhythm can then emerge from the SAN.*

8.6 The Worked Example: Analysing a VT Circuit

A 64-year-old man with a prior inferior MI has sustained monomorphic VT at 180 bpm. Electrophysiology study maps the re-entry circuit: circuit length $L = 8$ cm. Conduction velocity in the scar border zone $CV = 0.2$ m/s. Effective refractory period $ERP = 200$ ms.

Step 1 — Compute the wavelength. [Useful Approximation]

$$\lambda = CV \times ERP = 0.2 \times 0.2 = 0.04\ m = 4\ cm$$

Step 2 — Compute the excitation gap.

$$Gap = L - \lambda = 8 - 4 = 4\ cm$$

The central estimate of the gap is 4 cm; the range from propagated measurement uncertainty is 2.9–5.1 cm. This VT will be difficult to terminate with antiarrhythmic drugs alone — a large gap means the circuit has substantial redundancy, and drugs must dramatically increase λ to eliminate it.

Step 3 — Predict drug effect. *Amiodarone increases ERP from 200 to 320 ms:*

$$\lambda_{amio} = 0.2 \times 0.32 = 6.4\ cm$$

$$Gap_{amio} = 8 - 6.4 = 1.6\ cm$$

Amiodarone reduces the gap from 4 to 1.6 cm but does not eliminate it. The VT cycle length would slow (from 180 to L/CV_{new} bpm), but VT would persist.

Step 4 — Ablation target. *Catheter ablation must transect the circuit at its narrowest isthmus (the slow conduction zone in the scar). Eliminating the isthmus reduces the effective circuit length below λ, eliminating the gap. This is the mathematical basis of the ablation strategy: target the isthmus, not the entire circuit.*

> **Model Assumptions, Chapter 8.** *What the re-entry model assumes: (1) The circuit is a simple loop with uniform conduction velocity. (2) The wavelength $\lambda = CV \times ERP$ is spatially uniform. (3) The excitation gap is the only determinant of re-entry sustainability. Where it breaks down: Real arrhythmia circuits have heterogeneous CV and ERP (e.g. ischaemic scar border zones with complex fibrosis patterns). The wavelet hypothesis for AF is a statistical approximation — individual wavelets interact, collide, and fragment in ways the mean-field model does not capture. CV and ERP measurement uncertainty (±15–20%) propagates to a ±25–35% uncertainty in the excitation gap estimate.*

8.7 The Software Module

> **Module 8.1 — Arrhythmia Substrate Analyser**
>
> **Input:** *Circuit geometry (length L, number of limbs); conduction velocity (m/s) in each limb; effective refractory period (ms); atrial area (cm^2) for AF analysis; drug: effect on ERP and CV.*
>
> **Processing:**
>
> 1. *Compute wavelength $\lambda = CV \times ERP$.*
>
> 2. *Compute excitation gap (equation 8.2).*
>
> 3. *Classify: gap ≤ 0 (no re-entry), $0 < gap < 2$ cm (unstable re-entry), gap > 2 cm (stable re-entry).*
>
> 4. *Apply drug: recompute λ with modified ERP and CV; predict new gap and VT cycle length.*
>
> 5. *For AF: compute wavelet count (equation 8.3); predict cardioversion outcome.*
>
> 6. *EAD risk: compute inward/outward current ratio (equation 8.5) for given drug block.*
>
> **Output:** *Wavelength; excitation gap; re-entry classification; drug-*

> *modified gap; predicted VT cycle length; AF wavelet count; EAD risk index; ablation target recommendation.*
>
> *Full implementation at themathematicsoftheliving-body.com/modules.*

8.8 Chapter Summary

1. *Arrhythmias are bifurcations — qualitative changes in cardiac dynamical behaviour as a parameter crosses a critical threshold. Normal sinus rhythm is a stable limit cycle; re-entrant tachycardia is a different limit cycle; fibrillation is a chaotic attractor.*

2. *Re-entry requires three conditions: a circuit, unidirectional block, and path length $L > \lambda = CV \times ERP$. The excitation gap $L - \lambda$ determines whether re-entry is self-sustaining.*

3. *Class III drugs (increase ERP) increase λ and reduce the excitation gap. Class I drugs (reduce CV) paradoxically reduce λ and may enlarge the gap — proarrhythmic in structural heart disease.*

4. *AF requires a critical number of simultaneous wavelets $N \approx A/\lambda^2$. Cardioversion resets all wavelets to zero; ablation reduces A; chronic AF enlarges A and reduces λ, increasing N.*

5. *EADs arise from a Hopf bifurcation when the inward/outward current ratio exceeds 1 during Phase 3. Correction: avoid QT-prolonging drugs, correct hypokalaemia, use late I_{Na} blockers, pace faster.*

6. *In the VT worked example, the excitation gap was 4 cm. Amiodarone reduced it to 1.6 cm but did not eliminate it — predicting drug failure and the need for catheter ablation.*

Key Equations, Chapter 8

$$\lambda = CV \times ERP \quad \text{(cardiac wavelength)}$$

$$Gap = L - \lambda \quad \text{(excitation gap)}$$

$$N \approx \frac{A_{atrium}}{\lambda^2} \quad \text{(AF wavelet count)}$$

$$\frac{I_{Ca,L} + I_{Na,late}}{I_{Kr} + I_{Ks}} \gtrsim 1 \quad \Rightarrow \quad EAD \; risk$$

Key References, Chapter 8

1. *Moe, G.K. et al. (1964). A computer model of atrial fibrillation. Am. Heart J. 67, 200–220. [Original multiple wavelet hypothesis for AF.]*

2. *Mines, G.R. (1914). On circulating excitations in heart muscles and their possible relation to tachycardia and fibrillation. Trans. R. Soc. Can. 8, 43–52. [First description of re-entry circuit requirements.]*

3. *January, C.T. et al. (2014). 2014 AHA/ACC/HRS Guideline for the Management of Patients With Atrial Fibrillation. J. Am. Coll. Cardiol. 64, e1–e76. [Clinical translation of re-entry circuit theory.]*

Chapter 9

The Baroreflex: A Proportional-Integral Controller

9.1 The Physiology: Opening Part III

Parts I and II built the cardiac engine: the physics of pressure and flow, the mechanics of the ventricle, the electricity of the action potential, and the dynamical behaviour of the rhythm. Part III asks how the engine is regulated — how the cardiovascular system maintains stability under the constant disturbances of daily life: standing up, exercising, bleeding, breathing.

The primary regulator is the baroreflex — a negative feedback control loop that senses arterial pressure and adjusts heart rate and vascular resistance within seconds. It is the fastest of the cardiovascular control mechanisms (Chapter 1 introduced the multi-timescale hierarchy) and the one most accessible to mathematical analysis.

The baroreflex is not merely analogous to a feedback controller. It is a feedback controller, with an identifiable sensor, error signal, controller, and plant. Quantifying its gain, time constant, and set point from measurable data is possible — and clinically important, because baroreflex failure explains many phenomena that would otherwise seem inexplicable:

- *Why patients with autonomic neuropathy (diabetes, Parkinson's disease, spinal cord injury) develop dramatic blood pressure swings with posture changes*

- *Why bilateral carotid endarterectomy causes persistent hypertension*

- *Why athletes have resting bradycardia (enhanced baroreflex sensitivity)*

- *Why β-blockers reduce heart rate variability (they reduce the effector gain of the baroreflex)*

Before the equations: the molecular biology of the baroreceptor.

The baroreceptor is a mechanosensory neuron — a specialised primary afferent whose cell body lies in the nodose ganglion (aortic arch, cranial nerve X) or petrosal ganglion (carotid sinus, cranial nerve IX). Its peripheral terminal ending is embedded in the elastic wall of the arterial adventitia. When the wall is stretched by rising pressure, the terminal is deformed, opening mechanosensitive ion channels that depolarise the neuron.

Piezo2 as the primary baroreceptor transducer. *Recent work*

(Zeng et al., 2018) identified Piezo2 — a trimeric mechanosensitive cation channel distinct from the Piezo1 expressed in endothelium — as the primary transducer in baroreceptor afferents. Piezo2 knockout mice have severely impaired baroreflex function and labile blood pressure, confirming its essential role. Piezo2 opens within milliseconds of membrane stretch, allowing Na^+ and Ca^{2+} influx that triggers an action potential.

Central processing: the nucleus tractus solitarius. *Baroreceptor afferents synapse in the nucleus tractus solitarius (NTS) of the dorsal medulla. NTS glutamatergic neurons project to the caudal ventrolateral medulla (CVLM), which inhibits the rostral ventrolateral medulla (RVLM) — the pacemaker of sympathetic tone. High pressure $\to$ high baroreceptor firing $\to$ high NTS activity $\to$ high CVLM inhibition of RVLM $\to$ reduced sympathetic outflow $\to$ reduced HR and vasomotor tone. This is the neural implementation of the proportional term K_P in the PI control law.*

The integral term is real biology. *The K_I integral component is implemented by the dorsal vagal nucleus (DVN) and by humoral mechanisms: sustained low MAP releases vasopressin (AVP) from the posterior pituitary and activates the RAAS (Chapter 10), both of which accumulate over minutes to hours — exactly the integral of the error signal. The PI control framework provides a useful interpretive mapping onto this anatomy — approximate but illuminating.*

9.2 The Observation: The Baroreflex in Action

The classical demonstration of the baroreflex is the tilt-table test. A subject lying supine is tilted to 70° head-up over 5 seconds. Gravity shifts 500–700 mL of blood from the thorax to the lower extremities. Venous return falls. Cardiac output falls. Arterial pressure begins to fall.

Within 1–2 seconds, the baroreflex responds:

- *Heart rate increases (typically +10–20 bpm)*

- *Total peripheral resistance increases (arteriolar vasoconstriction)*

- *Blood pressure is restored within 15–30 s*

The entire sequence occurs without conscious awareness. In a healthy person, the pressure drop during orthostasis is < 10 mmHg systolic and < 5 mmHg diastolic. In baroreflex failure (autonomic dysfunction), the pressure drop can exceed 30 mmHg systolic — orthostatic hypotension.

Quantitatively, baroreflex sensitivity (BRS) is measured as the ratio of RR-interval change to systolic pressure change:

$$BRS = \frac{\Delta RR}{\Delta P_s} \quad [ms/mmHg] \tag{9.1}$$

Normal BRS is 10–20 ms/mmHg in healthy adults. BRS decreases with age, hypertension, and heart failure. $BRS < 3$ ms/mmHg indicates severely impaired baroreflex function.

9.3 The Mathematics: PI Control of Arterial Pressure

9.3.1 The Sensor: Baroreceptor Firing Rate

Baroreceptors are stretch-sensitive afferent neurons in the carotid sinus (cranial nerve IX) and aortic arch (cranial nerve X). Their firing rate f_{bar} increases with arterial pressure and decreases when pressure falls. The static characteristic is sigmoidal:

$$f_{bar}(P) = \frac{f_{\max}}{1 + e^{-k(P - P_{50})}} \tag{9.2}$$

where $f_{\max}$ is the maximum firing rate (≈ 100 Hz), k is the slope of the sigmoid (gain, ≈ 0.08 mmHg^{-1}), and P_{50} is the pressure at half-maximum firing (≈ 100 mmHg). For small deviations around the operating point, the sigmoid is approximately linear with slope (transduction gain):

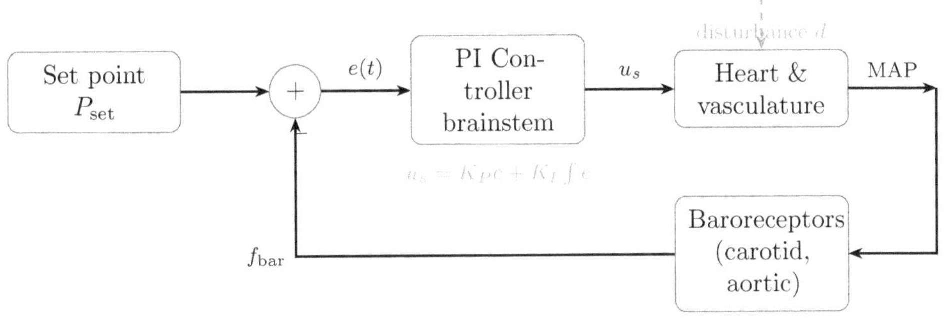

Figure 9.1: Baroreflex as a closed-loop proportional-integral (PI) control system. Baroreceptors (sensor) compare arterial MAP to the brainstem set point P_{set}, generating error $e(t)$. The brainstem (controller) applies PI control law $u_s = K_P e + K_I \int e$ dt, adjusting sympathetic outflow u_s to the heart and vasculature (plant). The integral term eliminates steady-state error — MAP returns exactly to P_{set} after any sustained disturbance.

$$G_s = k \cdot f_{\text{max}}/4 \approx 2 \ Hz/mmHg \tag{9.3}$$

9.3.2 The Controller: Proportional-Integral Structure

The brainstem cardiovascular centres (nucleus tractus solitarius, rostral ventrolateral medulla, dorsal vagal nucleus) process the baroreceptor signal and generate efferent autonomic output. This processing implements a proportional-integral (PI) control law.

Let $e(t) = P_{set} - P(t)$ be the pressure error. The efferent sympathetic output $u_s(t)$ is:

$$u_s(t) = K_P \cdot e(t) + K_I \int_0^t e(\tau) \, d\tau \tag{9.4}$$

The proportional term $K_P e(t)$ responds immediately to pressure deviations. The integral term $K_I \int e \, d\tau$ accumulates the error over time and eliminates steady-state offset — it ensures that MAP returns exactly to *the set point after a sustained disturbance, not merely close to it.*

In engineering terms: K_P provides speed of response; K_I provides accuracy. The baroreflex uses both.

9.3.3   The Plant: Heart and Vasculature

The plant (the system being controlled) consists of the heart and vasculature. The MAP response to sympathetic input u_s is modelled as a first-order system with time constant τ_p:

$$\tau_p \frac{\mathrm{d}P}{\mathrm{d}t} = -P + G_p u_s + d(t) \tag{9.5}$$

where G_p is the plant gain (mmHg per unit sympathetic output) and $d(t)$ is the disturbance (the orthostatic blood pressure drop, haemorrhage, etc.). Typical values: $\tau_p \approx 6$ s, $G_p \approx 3$ mmHg/Hz.

9.3.4   The Closed-Loop Response

Combining the PI controller and the plant, the closed-loop ODE for pressure is:

$$\tau_p \frac{\mathrm{d}P}{\mathrm{d}t} = -P + G_p G_s \left[K_P(P_{set} - P) + K_I \int_0^t (P_{set} - P)\,\mathrm{d}\tau \right] + d(t) \tag{9.6}$$

Differentiating to eliminate the integral:

$$\tau_p \ddot{P} + (1 + G_p G_s K_P)\dot{P} + G_p G_s K_I P = G_p G_s K_I P_{set} + \dot{d}(t) \tag{9.7}$$

This is a second-order linear ODE — the characteristic equation of a PI-controlled first-order plant. Its solutions are either overdamped (monotonic return to set point) or underdamped (oscillatory return). The condition for critical damping (fastest non-oscillatory response) is:

$$K_P = \frac{2\sqrt{K_I \tau_p / G_p G_s} - 1}{G_p G_s} \tag{9.8}$$

The Mayer waves — low-frequency oscillations in blood pressure at $\approx$ 0.1 Hz seen in continuous blood pressure recordings — are the underdamped oscillatory response of the baroreflex control loop. They are not noise. They are the ringing of the PI controller when its gain parameters are in the underdamped regime.

9.4 The Derivation: Baroreflex Gain from Transfer Function Analysis

The closed-loop transfer function of the baroreflex (from disturbance $D(s)$ to MAP $Y(s)$ in Laplace domain) is:

$$\frac{Y(s)}{D(s)} = \frac{\tau_p s}{\tau_p s^2 + (1 + G_p G_s K_P)s + G_p G_s K_I} \tag{9.9}$$

At steady state ($s \to 0$), the transfer function from disturbance to output approaches zero — the integral term eliminates steady-state error. This is the key advantage of the I-term: no matter how large the sustained disturbance, MAP returns exactly to P_{set}.

The open-loop gain *at DC (zero frequency) is:*

$$L_0 = G_p \cdot G_s \cdot K_P = 3 \times 2 \times K_P = 6K_P \tag{9.10}$$

For typical $K_P \approx 2$ Hz^{-1}: $L_0 \approx 12$. A loop gain of 12 means that the feedback reduces a disturbance by a factor of $1/(1 + L_0) \approx 1/13$ at the operating point. A 13 mmHg haemorrhagic pressure drop is corrected to ≈ 1 mmHg residual error at steady state.

9.4.1 Measuring Baroreflex Gain Clinically

BRS (equation 9.1) is the clinical measurement of the closed-loop baroreflex gain. It is measured by three methods:

Phenylephrine method. *Bolus IV phenylephrine raises MAP by 15–30 mmHg. The reflex bradycardia (RR interval prolongation) is measured. $BRS = \Delta RR/\Delta P_s$.*

Sequence method. *Beat-by-beat systolic pressure and RR interval are recorded. Sequences of ≥ 3 consecutive beats where both systolic pressure and RR interval change in the same direction (pressure up, RR longer; pressure down, RR shorter) are identified. The slope of RR vs. P_s in these sequences is BRS.*

Spectral method. *The transfer function gain between pressure and RR interval is computed in the low-frequency (0.04–0.15 Hz) and high-frequency (0.15–0.4 Hz) bands from spectral analysis of simultaneous pressure and RR recordings. This separates sympathetic (LF) and parasympathetic (HF) baroreflex contributions.*

9.5 The Clinical Interpretation: Baroreflex Failure and Autonomic Testing

9.5.1 Orthostatic Hypotension: Quantifying the Deficit

Orthostatic hypotension (OH) is defined as a drop in systolic BP $\geq$ 30 mmHg or diastolic BP $\geq$ 15 mmHg within 3 minutes of standing. From the control system model, OH occurs when:

$$\frac{1}{1 + L_0} \cdot \Delta d > 30 \ mmHg \tag{9.11}$$

where Δd is the orthostatic disturbance (typically 15–20 mmHg) and L_0 is the open-loop baroreflex gain. Rearranging:

$$L_0 < \frac{\Delta d}{30} - 1 \approx \frac{17}{30} - 1 < 0$$

For the normal baroreflex ($L_0 \approx 12$), the corrected pressure drop is $17/(1+12) \approx 1.3 \ mmHg$ — imperceptible. For baroreflex failure ($L_0 \to 0$), the *corrected drop equals the raw disturbance (≈ 17 mmHg) — below the OH threshold but symptomatic.*

For complete baroreflex failure ($L_0 = 0$), the uncorrected disturbance passes straight through: a 30 mmHg orthostatic drop, causing syncope.

9.5.2 Heart Rate Variability: A Window into Baroreflex Function

Heart rate variability (HRV) is the beat-to-beat variation in RR interval. It is dominated by two frequency components:

High frequency (HF, 0.15–0.4 Hz): *Respiratory sinus arrhythmia. During inspiration, intrathoracic pressure falls, venous return increases, MAP briefly rises, baroreceptors activate, and HR slows. During expiration, the reverse occurs. This HF component reflects parasympathetic (vagal) baroreflex gain.*

Low frequency (LF, 0.04–0.15 Hz): *Mayer waves. The ≈ 0.1 Hz oscillation from the underdamped baroreflex loop. This LF component reflects sympathetic baroreflex gain.*

The LF/HF ratio is used (controversially) as a measure of sympathovagal balance. Low HRV overall (reduced variability in both bands) is a robust predictor of post-MI mortality, independent of ejection fraction — because it reflects compromised baroreflex function and reduced ability to respond to haemodynamic disturbances.

__Why HRV predicts mortality after MI:__ Low HRV means low baroreflex gain (L_0 is small). Low L_0 means disturbances produce large, sustained pressure changes. In the post-MI period, when the myocardium is vulnerable to ischaemia and arrhythmia, large pressure swings increase myocardial oxygen demand, reduce coronary perfusion time, and increase the likelihood of a bifurcation into VF. The mathematics of control theory connects a simple clinical measurement (HRV) to a mechanistic pathway to sudden cardiac death.

9.6 The Worked Example: Quantifying Baroreflex Failure in Autonomic Neuropathy

A 58-year-old man with type 2 diabetes for 22 years undergoes autonomic testing. Lying supine: MAP 95 mmHg, HR 78 bpm. After tilt to 70°: MAP falls to 68 mmHg (drop 27 mmHg), HR rises to 84 bpm (increase 6 bpm). Phenylephrine test: MAP raised 22 mmHg, RR interval increased 40 ms.

Step 1 — Baroreflex sensitivity.

$$BRS = \frac{\Delta RR}{\Delta P_s} = \frac{40}{22} = 1.8 \ ms/mmHg \quad (normal \geq 10)$$

Severely reduced BRS. The baroreflex is operating at $< 20\%$ of normal sensitivity.

Step 2 — HR response to tilt. *Normal HR increase with tilt: 15–20 bpm. Observed: 6 bpm. The HR response is blunted — consistent with impaired sympathetic activation (autonomic neuropathy).*

Step 3 — Open-loop gain estimate. *From the tilt result: disturbance $d = 27$ mmHg, residual pressure drop $= 27$ mmHg (essentially uncorrected, since MAP fell from 95 to 68).*

$$L_0 \approx \frac{d}{\Delta P_{residual}} - 1 = \frac{27}{27} - 1 = 0$$

Open-loop gain ≈ 0: the baroreflex has essentially no corrective effect. The patient meets criteria for orthostatic hypotension (27 mmHg systolic drop) and has diabetic autonomic neuropathy with complete baroreflex impairment.

Step 4 — Management implications. *Pharmacological support: fludrocortisone (increases blood volume, raising the baseline MAP above which orthostatic drops become symptomatic) and midodrine (α_1 agonist, increases TPR without requiring an intact baroreflex). The goal is to shift* *the operating point, not to restore the lost feedback.*

Model Assumptions, Chapter 9. *What the PI baroreflex model assumes: (1) The baroreflex is a linear feedback controller around a fixed set point. (2) The plant (heart + vasculature) is a first-order system with time constant τ_p. (3) Sympathetic and parasympathetic limbs are not distinguished. Where it breaks down: The set point resets with chronic hypertension (baroreceptors adapt, violating the fixed set point assumption). The plant is nonlinear (sympathetic activation is bounded above). The two autonomic limbs have different dynamics: vagal responses occur within one heartbeat; sympathetic responses take 5–10 seconds. Treating them as a single PI controller conflates these timescales.*

9.7 The Software Module

Module 9.1 — Baroreflex Control System Simulator

Input: *Baroreflex sensitivity BRS (ms/mmHg); resting MAP and HR; orthostatic disturbance magnitude (mmHg); PI gains K_P and K_I (or estimate from BRS); plant time constant τ_p.*

Processing:

1. *Compute open-loop gain L_0 from BRS and plant parameters.*

2. *Solve the closed-loop ODE (equation 9.6) for a step disturbance using scipy.integrate.solve_ivp.*

3. *Compute steady-state pressure error and time to 90% correction.*

4. *Classify damping: overdamped, critically damped, underdamped (Mayer waves present).*

5. *Simulate Mayer waves: plot MAP oscillations at 0.1 Hz for underdamped gain parameters.*

6. *Compute LF and HF HRV power from simulated RR interval*

> *series.*
>
> **Output:** *MAP time course after orthostatic tilt; steady-state error; time to correction; BRS estimate; damping classification; Mayer wave simulation; LF/HF power.*
>
> *Full implementation at themathematicsoftheliving-body.com/modules.*

9.8   Chapter Summary

1. *The baroreflex is a PI controller for arterial pressure. Barorecep-tors are the sensor ($G_s \approx 2$ Hz/mmHg). The brainstem implements proportional-integral control (equation 9.4). The heart and vascula-ture are the plant ($G_p \approx 3$ mmHg/Hz, $\tau_p \approx 6$ s).*

2. *The integral term ensures zero steady-state error: MAP returns ex-actly to P_{set} after a sustained disturbance. This property distin-guishes PI control from proportional-only control.*

3. *The closed-loop ODE is second-order (equation 9.7). Its response is overdamped or underdamped depending on PI gains. Mayer waves (0.1 Hz blood pressure oscillations) are the underdamped ringing of the baroreflex loop.*

4. *Open-loop gain $L_0 = G_p G_s K_P \approx 12$ normally. A 17 mmHg ortho-static disturbance is corrected to < 1.4 mmHg residual error. When $L_0 \to 0$ (autonomic failure), the disturbance passes uncorrected.*

5. *BRS ($\Delta RR/\Delta P_s$, normal 10–20 ms/mmHg) is the clinical measure of baroreflex gain. Low HRV reflects low BRS and predicts post-MI mortality through the control theory mechanism.*

6. *In the diabetic autonomic neuropathy worked example, BRS was 1.8 ms/mmHg ($< 20\%$ of normal) and open-loop gain was ≈ 0. The orthostatic disturbance passed completely uncorrected, explaining the 27 mmHg pressure drop on tilt.*

Key Equations, Chapter 9

$$u_s(t) \;=\; K_P e(t) + K_I \int_0^t e(\tau)\,\mathrm{d}\tau \quad \textit{(PI controller)}$$

$$\lambda \;=\; CV \times ERP$$

$$BRS \;=\; \frac{\Delta RR}{\Delta P_s} \quad \textit{(baroreflex sensitivity)}$$

$$\frac{Y(s)}{D(s)} \;=\; \frac{\tau_p s}{\tau_p s^2 + (1 + G_p G_s K_P)s + G_p G_s K_I}$$

Key References, Chapter 9

1. *Eckberg, D.L. & Sleight, P. (1992). Human Baroreflexes in Health and Disease. Clarendon Press. [Comprehensive treatment of baroreflex physiology and clinical measurement.]*

2. *La Rovere, M.T. et al. (1998). Baroreflex sensitivity and heart-rate variability in prediction of total cardiac mortality after myocardial infarction. Lancet 351, 478–484. [Clinical validation of BRS as a mortality predictor.]*

3. *Zeng, W.Z. et al. (2018). PIEZOs mediate neuronal sensing of blood pressure and the baroreceptor reflex. Science 362, 464–467. [Molecular identification of Piezo2 as the baroreceptor transducer.]*

Chapter 10

The Renin-Angiotensin-Aldosterone System: A Hormonal Feedback Loop

The baroreflex corrects blood pressure in seconds. The RAAS corrects it over hours to days. Together they span six orders of magnitude in time — the fastest and slowest feedback loops in cardiovascular physiology. Understanding both is understanding blood pressure control.

— *Zachariah Sinkala*

10.1 The Physiology: A Slow but Powerful Second Line of Defence

Chapter 9 established the baroreflex as the fast feedback controller of arterial pressure — acting within seconds. But the baroreflex has a fundamental limitation: it adapts. Within 24–48 hours of a sustained pressure change, baroreceptors reset their firing threshold to the new pressure level. A chronically hypertensive patient's baroreceptors fire as if 160 mmHg were normal.

This adaptation means the baroreflex is unsuited for correcting chronic volume and pressure disturbances. A second system takes over for the slow timescale: the renin-angiotensin-aldosterone system (RAAS).

The RAAS is a hormonal cascade that regulates blood volume and pressure over hours to days. Its primary output is aldosterone, which acts on the kidney to retain sodium and water, expanding blood volume, raising venous return, and increasing MAP. When blood pressure falls, the cascade activates; when it rises, the cascade is suppressed. The loop closes through the kidney — long before MAP is fully restored, the RAAS acts on blood volume to bring it about.

The RAAS cascade has a precise kinetic structure: each step is an enzymatic reaction with a measurable rate constant. This makes it uniquely amenable to mathematical modelling — and to targeted drug intervention at every step.

> ***Before the equations: the proteins of the renin-angiotensin-aldosterone cascade.***
>
> *The RAAS is one of the most pharmacologically targeted pathways in all of medicine. Every step involves a named protein product, each of which has been the subject of drug development:*
>
> ***Renin*** *(encoded by REN) is an aspartyl protease synthesised as pre-prorenin in juxtaglomerular cells. It is stored in secretory granules and released by exocytosis on stimulation. Its substrate specificity is*

extremely narrow: it cleaves only the Leu-Val bond at position 10–11 of angiotensinogen, producing the 10-amino-acid angiotensin I.

Angiotensinogen *(encoded by AGT) is a serpin (serine protease inhibitor superfamily) produced constitutively by the liver and released into the circulation at $\approx 1\ \mu M$ — well above the K_m of renin ($\approx 1.6\ \mu M$), placing the reaction near saturation. Variants in AGT (e.g. M235T) are associated with essential hypertension, likely by increasing angiotensinogen plasma levels.*

ACE *(angiotensin-converting enzyme, encoded by ACE) is a zinc metalloprotease expressed on the luminal surface of pulmonary endothelial cells. It cleaves the C-terminal His-Leu dipeptide from angiotensin I to produce the 8-amino-acid angiotensin II. The same active site degrades bradykinin — which explains why ACE inhibitors cause cough and angioedema (bradykinin accumulation) and why ARBs (which block the AT_1 receptor, not ACE) do not.*

Aldosterone *is synthesised in the zona glomerulosa of the adrenal cortex from cholesterol via a steroidogenic pathway regulated by angiotensin II acting on AT_1 receptors. Aldosterone crosses cell membranes and binds the mineralocorticoid receptor (MR, encoded by NR3C2) in the nucleus, directly activating transcription of the ENaC α-subunit (SCNN1A) and the Na^+/K^+-ATPase α-subunit. This is why aldosterone effects are delayed by 1–2 hours (protein synthesis is required) and why mineralocorticoid excess causes hypokalaemia (K^+ lost in exchange for Na^+ retained).*

10.2 The Observation: The Three-Step Cascade

The RAAS proceeds through three sequential enzymatic steps:

Step 1 — Renin release. *Renin is a protease enzyme secreted by the juxtaglomerular (JG) cells of the kidney in response to three stimuli:*

- *Reduced renal perfusion pressure (afferent arteriolar stretch receptor)*

- *Reduced Na^+/Cl^- delivery to the macula densa (tubuloglomerular feedback)*

- *Sympathetic activation (β_1-receptor on JG cells)*

Step 2 — Angiotensin II generation. *Renin cleaves angiotensinogen (produced by the liver) to angiotensin I. Angiotensin-converting enzyme (ACE) in the lung converts angiotensin I to angiotensin II. Angiotensin II is the primary effector molecule of the RAAS.*

Step 3 — Aldosterone secretion. *Angiotensin II acts on the adrenal cortex (zona glomerulosa) to stimulate aldosterone synthesis and release. Aldosterone acts on the cortical collecting duct of the kidney to increase Na^+ reabsorption (via ENaC channels) and K^+ excretion. Water follows sodium, expanding extracellular volume and blood volume.*

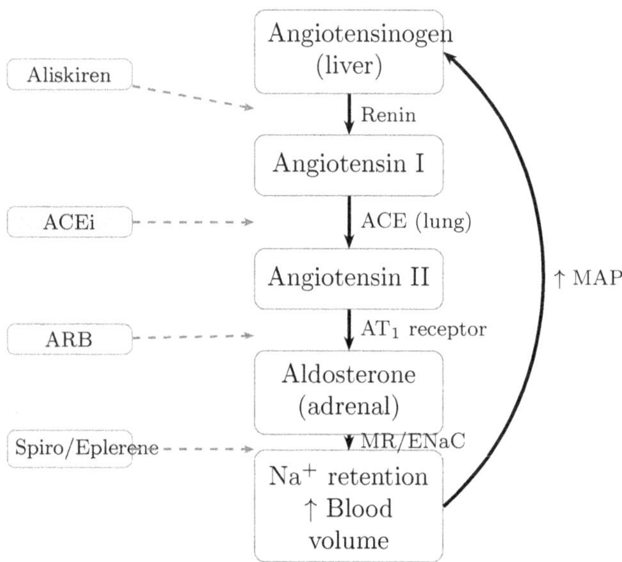

Figure 10.1: The RAAS cascade and drug targets. Each enzymatic step (right labels) has a corresponding drug class (blue, left) that inhibits it. ACEi and ARBs block adjacent steps but via different mechanisms: ACEi reduces AngII production (and spares bradykinin degradation), while ARBs block AngII action at the AT_1 receptor (no bradykinin effect). The feedback arrow shows the long-term pressure-volume homeostasis loop.

10.3 The Mathematics: Enzyme Kinetics at Each Step

10.3.1 Michaelis-Menten Kinetics for Each Enzymatic Step

Each step in the RAAS cascade is an enzymatic reaction. For an enzyme E acting on substrate S to produce product P, the Michaelis-Menten rate law gives the reaction velocity:

[Core Principle]

$$v = \frac{V_{\max}[S]}{K_m + [S]} \tag{10.1}$$

where $V_{\max}$ is the maximum reaction velocity (proportional to enzyme concentration) and K_m is the Michaelis constant (substrate concentration for half-maximum velocity).

Step 1 — Renin-angiotensinogen reaction:

$$v_1 = \frac{V_{\max,R} \cdot [AGT]}{K_{m,R} + [AGT]} \cdot f_R(P, NaCl, SNS) \tag{10.2}$$

where $[AGT]$ is angiotensinogen concentration, $K_{m,R} \approx 1.6~\mu M$ is the Michaelis constant for renin acting on angiotensinogen, and f_R is the regulatory function capturing the three stimuli for renin release:

$$f_R = f_P(P_{renal}) \cdot f_{NaCl} \cdot f_{SNS}(\beta_1) \tag{10.3}$$

Step 2 — ACE reaction (AngI to AngII):

$$v_2 = \frac{V_{\max,ACE} \cdot [AngI]}{K_{m,ACE} + [AngI]} \cdot (1 - I_{ACEi}) \tag{10.4}$$

where $I_{ACEi} \in [0,1]$ is the fractional inhibition by an ACE inhibitor drug. When $I_{ACEi} = 0.9$ (90% inhibition, a typical therapeutic ACE inhibitor dose), $V_{\max,ACE}$ is effectively reduced by 90%, dramatically re-

ducing AngII production.

Step 3 — Aldosterone production:

$$\frac{\mathrm{d}[Aldo]}{\mathrm{d}t} = k_{prod} \cdot [AngII]^n - k_{deg} \cdot [Aldo] \qquad (10.5)$$

where $n \approx 1.5$ is the cooperativity of AngII stimulation of the adrenal cortex and $k_{deg} \approx 0.017 \ min^{-1}$ is the aldosterone degradation rate (half-life $\approx 40 \ min$).

10.3.2   The Full RAAS ODE System

Combining all three steps with conservation of mass:

$$\frac{\mathrm{d}[Renin]}{\mathrm{d}t} = S_R(P) - k_R[Renin] \qquad (10.6)$$

$$\frac{\mathrm{d}[AngI]}{\mathrm{d}t} = v_1 - v_2 - k_{AngI}[AngI] \qquad (10.7)$$

$$\frac{\mathrm{d}[AngII]}{\mathrm{d}t} = v_2 - k_{AngII}[AngII] \qquad (10.8)$$

$$\frac{\mathrm{d}[Aldo]}{\mathrm{d}t} = k_{prod}[AngII]^n - k_{deg}[Aldo] \qquad (10.9)$$

$$\frac{\mathrm{d}V_b}{\mathrm{d}t} = k_{Na}[Aldo] - k_{loss}V_b \qquad (10.10)$$

where V_b is blood volume, $S_R(P)$ is the pressure-dependent renin secretion rate, and k_{Na} converts aldosterone activity to sodium (and thus volume) retention.

The time constants of this system span three decades: renin has a half-life of $\approx 15 \ min$, AngI $\approx 30 \ s$, AngII $\approx 30 \ s$, aldosterone $\approx 40 \ min$, and blood volume changes occur over hours to days. This separation of timescales justifies a quasi-steady-state analysis for the fast variables (AngI, AngII) while treating blood volume as the slow variable.

10.4 The Derivation: Why Salt Restriction Lowers Blood Pressure

The long-term blood pressure set point is determined by the pressure-natriuresis curve — the relationship between MAP and renal sodium excretion. At steady state, sodium intake must equal sodium excretion.

Guyton's model of long-term pressure control states: the kidney is the ultimate determinant of long-term MAP because it sets the operating point of the pressure-natriuresis curve.

The steady-state blood volume V_b^* satisfies:

$$Na_{intake}^+ = Na_{excretion}^+(P^*, V_b^*) \tag{10.11}$$

The RAAS modifies the pressure-natriuresis curve by shifting the setpoint. When RAAS is activated (low pressure), aldosterone retention shifts the curve rightward — the kidney retains more sodium at any given pressure, raising V_b and raising MAP until the new balance is reached.

Salt restriction lowers MAP because:

1. *Reduced Na^+ intake lowers V_b acutely*

2. *Reduced V_b reduces venous return, reducing CO, reducing MAP*

3. *Reduced MAP activates renin release*

4. *But: at the new lower intake, the RAAS can only restore V_b partially before sodium balance is re-established at a lower MAP*

The mathematical prediction: a reduction in Na^+ intake from 200 to 100 mEq/day shifts MAP by approximately:

*[**Approximation** — see **Model Assumptions**]*

$$\Delta MAP \approx -\frac{\Delta Na_{intake}^+}{d\,Na_{exc}^+/dP} \tag{10.12}$$

where the denominator is the slope of the pressure-natriuresis curve

(≈ 3 mEq/day per mmHg in salt-sensitive hypertension). For $\Delta Na^+ = -100$ mEq/day:

$$\Delta MAP \approx -\frac{100}{3} \approx -33 \ mmHg$$

This is the upper bound of salt restriction efficacy in salt-sensitive hypertension. The clinical literature shows 5–10 mmHg reductions — smaller because real humans are not perfectly salt-sensitive and have other volume regulators. But the mathematical framework correctly predicts the direction and approximate magnitude.

10.5 The Clinical Interpretation: Targeting the RAAS Cascade

The RAAS is the most drug-targeted system in cardiovascular medicine. Every step of the cascade has a corresponding drug class:

Target	Drug class	Mechanism	ΔP (mmHg)
Renin	*Aliskiren*	*Blocks renin activity*	*−8*
ACE	*ACEi*	*Reduces AngI→AngII*	*−10*
AT_1R	*ARB*	*Blocks AngII receptor*	*−10*
Aldo	*Spiro/Epleren.*	*Blocks aldo. receptor*	*−5*
ENaC	*Amiloride*	*Blocks Na^+ channel*	*−4*

The MAP reductions are additive only when drugs target non-overlapping cascade steps. ACEi and ARB block adjacent steps but have overlapping effects — combining them does not double the MAP reduction and increases the risk of hyperkalaemia and renal failure (dual RAAS blockade is not recommended).

10.5.1 ACE Inhibitors: Blocking the Key Step

*ACE inhibitors reduce AngII production by $\approx$ 60–80% at therapeutic doses.
From equation (10.4) with $I_{ACEi} = 0.75$:*

$$v_{2,treated} = 0.25 \times v_{2,normal}$$

*AngII falls. Aldosterone falls. Sodium excretion increases. Blood volume
falls. MAP falls. But compensatory renin release increases (the feedback
loop is not completely broken): renin and AngI rise, limiting the AngII
reduction to $\approx$ 60–70% at steady state rather than the 75% implied by
the single-step block. This* ACE escape *is why ARBs (blocking the AngII
receptor rather than its production) may provide more complete RAAS
blockade.*

10.5.2 Heart Failure and RAAS Activation

*In heart failure, reduced CO reduces renal perfusion pressure. The kid-
ney interprets this as hypovolaemia and activates the RAAS. Renin rises.
AngII rises. Aldosterone rises. Sodium is retained. Blood volume increases.*

*This is appropriate in true hypovolaemia but catastrophically counterpro-
ductive in heart failure: the increased blood volume raises filling pressures
(worsening pulmonary oedema) without improving cardiac output (because
the failing ventricle is operating at the plateau of its Frank-Starling curve).
The RAAS activation that is meant to restore perfusion pressure instead
floods the lungs.*

*This is the mechanistic rationale for ACE inhibitors and ARBs in heart
failure: block the RAAS, prevent the maladaptive volume retention, reduce
preload and afterload, improve survival.*

Why spironolactone reduces mortality in heart failure: *Aldosterone does more than retain sodium. At the high levels seen in
severe heart failure, aldosterone causes myocardial fibrosis (collagen
deposition in the cardiac interstitium), reduces vascular compliance,*

and promotes ventricular remodelling. Blocking the aldosterone receptor (with spironolactone or eplerenone) reduces mortality by 30% in severe heart failure — not primarily through its diuretic effect but through its anti-fibrotic and anti-remodelling effects. The enzyme kinetics of equation (10.9) predict which patients have the highest aldosterone levels (lowest CO, highest RAAS activation) and therefore the greatest benefit from aldosterone receptor blockade.

10.6 The Worked Example: Modelling ACE Inhibitor Response in Hypertension

A 52-year-old woman with essential hypertension (MAP = 108 mmHg, normal 93 mmHg, on no treatment) is started on lisinopril 10 mg daily. Her plasma renin activity (PRA) at baseline is 1.8 ng/mL/h (normal 0.5–2.0). Serum aldosterone 18 ng/dL (normal 3–16). Serum K^+ 4.1 mEq/L.

Step 1 — Characterise the RAAS at baseline. *PRA and aldosterone are mildly elevated, consistent with renin-mediated hypertension (normal renin activity but high for the degree of hypertension — the RAAS is not fully suppressed as it should be at MAP 108 mmHg). The aldosterone-renin ratio (ARR):*

$$ARR = \frac{18}{1.8} = 10 \ ng/dL \ per \ ng/mL/h \quad (normal < 20)$$

ARR < 20: primary aldosteronism is unlikely. This is renin-dependent hypertension.

Step 2 — Predict ACE inhibitor response. *Lisinopril at 10 mg achieves ≈ 70% ACE inhibition at steady state. From the ODE system, AngII will fall by ≈ 60% at steady state (accounting for compensatory renin rise). Aldosterone will fall by ≈ 55%. From the pressure-volume model, the predicted MAP reduction:*

$$\Delta MAP \approx -0.55 \times \frac{[Aldo]_0}{[Aldo]_0/MAP_0} \approx -8 \ to \ -12 \ mmHg$$

Target MAP: 93–96 mmHg. Predicted MAP on lisinopril: 96–100 mmHg. A second agent (thiazide diuretic or calcium channel blocker) will likely be needed for full control.

Step 3 — Monitor potassium. *ACE inhibitors reduce aldosterone, reducing K^+ excretion. Predicted K^+ rise: 0.2–0.5 mEq/L. Expected post-treatment K^+: 4.3–4.6 mEq/L. Recheck K^+ and creatinine at 1–2 weeks.*

Step 4 — Identify responders vs. non-responders. *High-renin hypertension (PRA > 1.5 ng/mL/h) responds well to ACE inhibitors and ARBs. Low-renin hypertension (PRA < 0.5 ng/mL/h, as in primary aldosteronism or volume-dependent hypertension) responds poorly to RAAS blockade and better to diuretics or calcium channel blockers. The RAAS ODE system predicts this: when renin is suppressed, blocking ACE has little effect because AngII production is already minimal.*

Model Assumptions, Chapter 10. *What the RAAS ODE model assumes: (1) Each enzymatic step follows Michaelis-Menten kinetics with fixed K_m and $V_{\max}$. (2) Angiotensin II acts only via the AT_1 receptor. (3) Blood volume changes are proportional to sodium retention. Where it breaks down: ACE escape: ACE inhibitors cause a compensatory rise in renin that partially restores AngII. The model predicts this qualitatively but the magnitude of ACE escape varies 2-fold between individuals. The RAAS interacts with the sympathetic system (SNS stimulates renin release) and with potassium homeostasis (aldosterone raises K^+ excretion) in ways not fully captured by the five-ODE system.*

10.7   The Software Module

Module 10.1 — RAAS Cascade ODE Simulator

Input: *Baseline MAP, PRA, aldosterone, serum K^+; drug: ACEi inhibition fraction, ARB block fraction, aldosterone antagonist block fraction; Na^+ intake (mEq/day); simulation duration (hours).*

Processing:

1. *Initialise RAAS ODE system (equations 10.6–10.10) at baseline steady state.*

2. *Apply drug: modify ACE V_{max} (equation 10.4), AngII receptor occupancy, or aldosterone receptor block.*

3. *Integrate ODE system using scipy.integrate.solve_ivp (stiff, LSODA).*

4. *Compute quasi-steady-state for AngI and AngII; integrate blood volume ODE.*

5. *Predict MAP from blood volume change using the pressure-volume relationship.*

6. *Compute expected K^+ change from aldosterone reduction.*

Output: *Time courses of renin, AngI, AngII, aldosterone, blood volume, MAP, K^+; steady-state predictions; drug response classification (renin-dependent vs. volume-dependent); combination therapy prediction.*

Full implementation at themathematicsoftheliving-body.com/modules.

10.8   Chapter Summary

1. *The RAAS operates over hours to days, complementing the second-timescale baroreflex. Together they span six orders of magnitude in response time.*

2. *Each step of the cascade (renin, ACE, aldosterone) follows Michaelis-Menten kinetics. The full RAAS is a five-ODE system with time constants from 30 s (AngII) to days (blood volume).*

3. *Salt restriction lowers MAP by shifting the pressure-natriuresis balance: $\Delta MAP \approx -\Delta Na^+_{intake}/(d\,Na^+_{exc}/dP)$. In salt-sensitive hypertension, 100 mEq/day reduction predicts ≈ 33 mmHg reduction; clin-*

ical reality is 5–10 mmHg.

4. *ACE inhibitors block the key enzymatic step (AngI $\to$ AngII) with $\approx$ 70% inhibition at therapeutic doses. Compensatory renin rise limits AngII reduction to $\approx$ 60% at steady state (ACE escape).*

5. *In heart failure, RAAS activation is maladaptive: it retains sodium to restore perceived hypovolaemia but raises preload on a failing ventricle already at its Frank-Starling plateau, worsening oedema without improving CO.*

6. *Spironolactone reduces mortality in heart failure through anti-fibrotic and anti-remodelling effects of aldosterone blockade, not primarily through diuresis. The ODE system predicts which patients have the highest aldosterone and therefore the greatest benefit.*

Key Equations, Chapter 10

$$v = \frac{V_{\max}[S]}{K_m + [S]} \quad \text{(Michaelis-Menten)}$$

$$\frac{\mathrm{d}[Aldo]}{\mathrm{dt}} = k_{prod}[AngII]^n - k_{deg}[Aldo]$$

$$\Delta MAP \approx -\frac{\Delta Na^+_{intake}}{\mathrm{d}\,Na^+_{exc}/\,\mathrm{d}P} \quad \text{(salt-pressure)}$$

$$v_{2,ACEi} = (1 - I_{ACEi}) \cdot v_{2,normal}$$

Key References, Chapter 10

1. *Guyton, A.C. (1980). Arterial pressure and hypertension. Circ. Res. 46 (Suppl I), I-2–I-12. [The pressure-natriuresis model of long-term BP control.]*

2. *CONSENSUS Trial Study Group (1987). Effects of enalapril on mortality in severe congestive heart failure. N. Engl. J. Med. 316, 1429–1435. [First demonstration of ACEi survival benefit in HF.]*

3. Pitt, B. et al. (1999). *The effect of spironolactone on morbidity and mortality in patients with severe heart failure. N. Engl. J. Med.* **341**, 709–717. [RALES trial — aldosterone antagonism in heart failure.]

144

Chapter 11

Autoregulation: How Vessels Think Locally

11.1   The Physiology: Local Flow Control Without Central Command

*Chapters 9 and 10 described centralised blood pressure regulation: the
baroreflex (brainstem) and the RAAS (kidney) both regulate systemic
MAP through central mechanisms.   But most organs do not wait for central instructions.   They regulate their own blood flow locally — a property*

called autoregulation.

Autoregulation is the ability of an organ's vasculature to maintain nearly constant blood flow over a wide range of perfusion pressures. When pressure rises, local arterioles constrict. When pressure falls, they dilate. The result is a plateau in the flow-pressure relationship — the autoregulation curve.

Three organs have the most robust autoregulation:

- **Brain:** *Flow maintained between MAP 60–150 mmHg. Below 60: ischaemia. Above 150: breakthrough hyperperfusion (hypertensive encephalopathy).*

- **Kidney:** *Flow maintained between MAP 80–180 mmHg, protecting glomerular filtration rate.*

- **Heart:** *Coronary flow maintained between coronary perfusion pressures of 60–130 mmHg.*

Understanding autoregulation mathematically is essential because its failure explains several clinical catastrophes:

- *Hypertensive encephalopathy when MAP exceeds the upper limit of cerebral autoregulation*

- *Contrast nephropathy in patients with impaired renal autoregulation*

- *Perioperative stroke when MAP is allowed to fall below the cerebral autoregulation lower limit*

- *Why rapidly lowering blood pressure in an acute hypertensive emergency can cause a stroke — the autoregulation curve has shifted rightward in chronic hypertensives*

11.2 The Observation: The Autoregulation Curve

The autoregulation curve plots organ blood flow against perfusion pressure. It has three regions:

Below the lower limit of autoregulation (LLA): *Flow falls linearly with pressure (passive vessels, maximal dilation already).*

Autoregulation plateau (LLA to ULA): *Flow is nearly constant despite changing pressure. Arteriolar resistance adjusts to maintain flow.*

Above the upper limit of autoregulation (ULA): *Flow rises linearly with pressure (forced vasodilation, vessels cannot constrict further).*

For the cerebral circulation: LLA $\approx$ 60 mmHg, ULA $\approx$ 150 mmHg (normal MAP range of autoregulation = 90 mmHg wide).

Critically, in chronic hypertension, the entire curve shifts rightward: LLA rises to $\approx$ 90 mmHg and ULA rises to $\approx$ 180 mmHg. A hypertensive patient is protected from high pressures but vulnerable to normal pressures: dropping their MAP from 160 to 100 mmHg — a clinically "safe" reduction — may push them below their shifted LLA.

11.3 The Mathematics: The Myogenic Response as a Feedback Loop

11.3.1 Three Mechanisms of Autoregulation

Three mechanisms contribute to autoregulation in different proportions depending on the organ:

Mechanism 1 — Myogenic response. *Arteriolar smooth muscle cells sense wall tension (via stretch-activated Ca^{2+} channels) and contract when stretched (increased pressure) or relax when compressed (reduced pressure). This is the fastest mechanism ($\tau \approx$ 5–15 s).*

Mechanism 2 — Metabolic regulation. *Metabolic byproducts of tissue activity (CO_2, H^+, K^+, adenosine) cause local vasodilation when accumulate. Higher metabolic rate $\rightarrow$ more byproducts $\rightarrow$ more dilation $\rightarrow$ more flow $\rightarrow$ washout of byproducts. This dominates in the coronary circulation.*

Mechanism 3 — Tubuloglomerular feedback. *In the kidney, NaCl delivery to the macula densa regulates afferent arteriolar tone, controlling* *glomerular filtration rate. This is organ-specific.*

11.3.2   The Myogenic Response ODE

Model the arteriolar radius $r(t)$ as dynamically regulated by the transmural pressure P_{tm} through the myogenic response. The equilibrium radius $r_\infty(P_{tm})$ decreases with increasing pressure (constriction):

$$r_\infty(P_{tm}) \;=\; r_0 \cdot \frac{1}{1 + k_m(P_{tm} - P_0)} \tag{11.1}$$

where r_0 is the passive radius at reference pressure P_0, and k_m is the myogenic gain (slope of the pressure-radius relationship, ≈ 0.01 $mmHg^{-1}$). The dynamic approach to equilibrium:

[Useful Approximation]

$$\tau_m \frac{\mathrm{d}r}{\mathrm{d}t} \;=\; r_\infty(P_{tm}) - r \tag{11.2}$$

where $\tau_m \approx 10$ s is the myogenic time constant.

11.3.3   Flow-Pressure Relationship from Poiseuille and Myogenic Response

From Chapter 2, flow through the arteriole: $Q = \pi\Delta P r^4/8\eta L$. At steady state ($\mathrm{d}r/\mathrm{d}t = 0$), $r = r_\infty(P_{tm})$. Substituting:

$$\begin{aligned}
Q(P) \;&=\; \frac{\pi\Delta P}{8\eta L} \cdot r_\infty(P)^4 \\[2mm]
&=\; \frac{\pi\Delta P}{8\eta L} \cdot \frac{r_0^4}{[1 + k_m(P - P_0)]^4}
\end{aligned} \tag{11.3}$$

This is the autoregulation curve from first principles.

As P increases: the numerator (ΔP) increases, tending to increase Q; the denominator $[1 + k_m(P - P_0)]^4$ also increases (constriction), tending to decrease Q. The plateau occurs when these two effects exactly cancel:

$$\frac{\mathrm{d}Q}{\mathrm{d}P} = 0 \Leftrightarrow \frac{1}{\Delta P} = \frac{4k_m}{1 + k_m(P - P_0)} \qquad (11.4)$$

Solving gives the pressure range over which autoregulation maintains constant flow.

11.3.4   The Autoregulation Index

The autoregulation index (AI) quantifies how effectively flow is maintained:

[Useful Approximation]

$$AI = 1 - \frac{\Delta Q/Q}{\Delta P/P} = 1 - \frac{dQ/Q}{dP/P} \qquad (11.5)$$

$AI = 1$: perfect autoregulation (flow completely independent of pressure). $AI = 0$: no autoregulation (passive vessels, flow proportional to pressure). Normal cerebral $AI \approx 0.8$–0.95.

11.4   The Derivation: Why Autoregulation Has an Upper Limit

The upper limit of autoregulation (ULA) arises from the maximum active constriction capacity of the vascular smooth muscle. Define $r_{\min}$ as the minimum radius achievable (maximum constriction). The ULA is the pressure at which $r_\infty(P_{ULA}) = r_{\min}$:

$$P_{ULA} = P_0 + \frac{1}{k_m}\left(\frac{r_0}{r_{\min}} - 1\right) \qquad (11.6)$$

For typical values: $r_0 = 50\ \mu m$, $r_{\min} = 25\ \mu m$ (maximum 50% constriction), $k_m = 0.01\ mmHg^{-1}$, $P_0 = 100\ mmHg$:

$$P_{ULA} = 100 + \frac{1}{0.01}\left(\frac{50}{25} - 1\right) = 100 + 100 = 200\ mmHg$$

But smooth muscle cannot sustain maximum contraction indefinitely. At very high pressures, the physical distending force exceeds the active contractile force — vessels are forcibly dilated (breakthrough perfusion). This sets the true ULA at ≈ 150 mmHg for cerebral vessels, below the theoretical maximum.

Similarly, the lower limit of autoregulation (LLA) arises from the maximum vasodilation capacity: $r_{\max}$ is the fully dilated radius:

$$P_{LLA} = P_0 - \frac{1}{k_m}\left(1 - \frac{r_0}{r_{\max}}\right) \tag{11.7}$$

Below P_{LLA}, the arteriole cannot dilate further, flow falls passively with pressure, and ischaemia results.

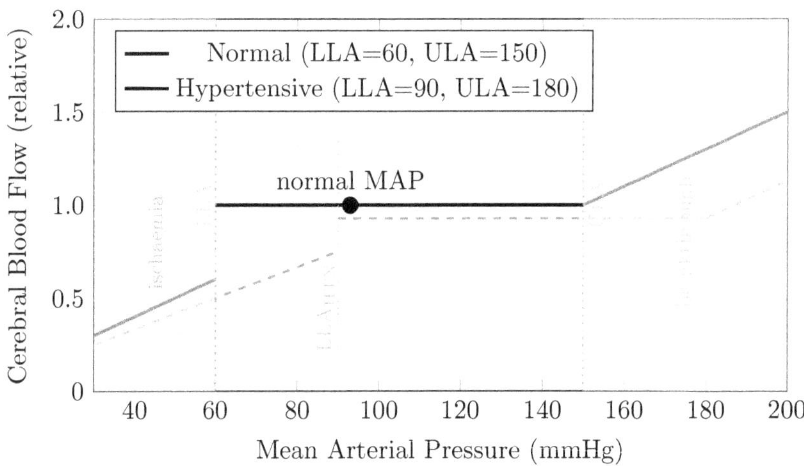

Figure 11.1: Cerebral autoregulation curves. The normal brain (black) maintains constant blood flow between LLA 60 mmHg and ULA 150 mmHg. Below the LLA (red zone): vessels are maximally dilated, flow falls passively — ischaemia. Above the ULA (orange zone): vessels are forcibly dilated, flow rises passively — hypertensive encephalopathy. Chronic hypertension shifts the entire curve rightward (red dashed): LLA rises to ≈ 90 mmHg, making "normal" MAP dangerous for these patients.

11.5 The Clinical Interpretation: When Autoregulation Fails

11.5.1 Hypertensive Emergency: The Rightward Shift

Chronic hypertension causes vascular remodelling: smooth muscle hypertrophies, the vessel wall thickens, and the myogenic gain k_m increases. The autoregulation curve shifts rightward: LLA rises from 60 to 90 mmHg.

This means the chronically hypertensive brain cannot tolerate a "normal" MAP. Rapidly reducing MAP from 180 to 120 mmHg (still hypertensive by normal standards) may cause cerebral ischaemia if the LLA has shifted above 120 mmHg.

The clinical rule from the mathematics: *In a hypertensive emergency, do not reduce MAP by more than 20–25% in the first hour. This is not arbitrary caution — it is the mathematical consequence of the rightward-shifted autoregulation curve.*

11.5.2 Ischaemic Stroke: Lost Autoregulation

In the ischaemic penumbra (tissue surrounding a stroke core), autoregulation is impaired. The AI falls toward zero. Flow becomes pressure-passive. In this setting:

- *Hypotension reduces flow to already-ischaemic tissue, extending the infarct*

- *Hypertension (which the body induces reflexively to maintain perfusion) may actually increase penumbral flow*

This is why permissive hypertension (allowing MAP 140–180 mmHg) is standard in acute ischaemic stroke management — preserving penumbral flow while autoregulation is absent.

11.5.3 Cerebral Perfusion Pressure in Raised ICP

In patients with raised intracranial pressure (ICP), the effective perfusion pressure for the brain is not MAP but cerebral perfusion pressure (CPP):

[Core Principle]

$$CPP \ = \ MAP - ICP \tag{11.8}$$

Normal ICP is < 15 mmHg, giving $CPP \approx MAP - 15$. When ICP rises (trauma, hydrocephalus), CPP falls even if MAP is normal. Autoregulation attempts to maintain cerebral blood flow by dilating arterioles, but if CPP falls below the LLA (≈ 50 mmHg in this context), ischaemia results.

Target CPP in severe traumatic brain injury: 50–70 mmHg. At $CPP < 50$ mmHg: critically impaired cerebral perfusion. This threshold is derived from the LLA of the cerebral autoregulation curve.

The Cushing reflex — autoregulation's last resort: *When ICP rises to the point where CPP is critically reduced, the brainstem detects ischaemia and generates a massive sympathetic response: severe hypertension, bradycardia, and irregular respirations (the Cushing triad). This is the body attempting to restore CPP by raising MAP above the ICP. It is a sign of impending brainstem herniation. The mathematics: when $CPP = MAP - ICP$ approaches zero, the body raises MAP to its maximum (≈ 200 mmHg) to keep CPP above the LLA. The Cushing reflex is the autoregulation ODE driving the baroreflex set point to its maximum.*

11.6 Metabolic Autoregulation: The Coronary Circulation

The coronary circulation relies more heavily on metabolic autoregulation than on the myogenic response. The heart's oxygen extraction is already near maximal at rest ($\approx 70\%$), so it cannot increase extraction when demand rises — it must increase flow. The mechanism:

1. *Increased myocardial work $\rightarrow$ increased O_2 consumption $\rightarrow$ decreased O_2 tension*

2. *Hypoxia activates K_{ATP} channels in coronary smooth muscle $\rightarrow$ hyperpolarisation $\rightarrow$ vasodilation*

3. *Adenosine (from ATP breakdown) released $\rightarrow$ A_2 receptor activation $\rightarrow$ coronary vasodilation*

4. *Increased flow $\rightarrow$ restored O_2 delivery $\rightarrow$ closed loop*

The metabolic vasodilation is quantified by the coronary flow reserve (CFR):

$$CFR = \frac{Q_{max}}{Q_{rest}} \qquad (11.9)$$

Normal $CFR \approx 3\text{--}5$: coronary flow can increase 3–5-fold from rest to maximum demand. $CFR < 2$ indicates impaired coronary autoregulation — coronary microvascular disease or significant epicardial stenosis. This is measured clinically by PET scanning or fractional flow reserve (FFR, Chapter 2).

11.7 The Worked Example: Calculating the Safe MAP Range After Stroke

A 71-year-old man presents with acute right MCA ischaemic stroke. He has a 20-year history of hypertension, with a pre-stroke baseline MAP of approximately 130 mmHg. On arrival, MAP is 165 mmHg.

Step 1 — Estimate shifted autoregulation limits. *In chronic hypertension with baseline MAP 130 mmHg, the autoregulation curve is shifted rightward. Estimated LLA:*

$$LLA_{shifted} \approx LLA_{normal} + 0.55 \times (MAP_{chronic} - 93)$$
$$= 60 + 0.55 \times 37 \approx 80 \; mmHg$$

(The 0.55 factor reflects the incomplete rightward shift, as the autoregulation curve does not move as far as the chronic pressure elevation.)

Step 2 — Assess the current MAP. *MAP 165 mmHg: well above the shifted ULA for most hypertensives (≈ 160 mmHg). This patient is near or above his ULA — hypertensive encephalopathy risk is present but the stroke itself may be causing the pressure elevation (Cushing-like response to ischaemia).*

Step 3 — Safe MAP reduction. *Standard recommendation: reduce MAP by no more than 15% in the first 24 hours in acute stroke (unless MAP > 220/120 mmHg).*

$$MAP_{target} = 165 \times (1 - 0.15) \approx 140 \ mmHg$$

Target MAP of 140 mmHg. This keeps him above the estimated LLA (≈ 80 mmHg) with a large safety margin while reducing the risk of haemorrhagic transformation.

Step 4 — If thrombolysis is given. *With IV tPA, MAP must be kept $\leq 180/105$ mmHg. Target 140–170 mmHg. The mathematics of the autoregulation curve defines both the upper bound (stroke risk if too high) and the lower bound (ischaemia extension if too low) for the safe treatment window.*

Model Assumptions, Chapter 11. *What the autoregulation model assumes: (1) Autoregulation is mediated solely by the myogenic response (a single ODE). (2) The steady-state radius function $r_{\infty}(P)$ is monotonically decreasing (no oscillatory behaviour). (3) The autoregulation curve is symmetric and continuous. Where it breaks down: Real cerebral autoregulation includes metabolic, neurogenic, and endothelial components in addition to the myogenic response. The autoregulation curve has substantial inter-individual variability (± 15 mmHg in the LLA). In acute brain injury, autoregulation may be absent, impaired, or shifted without warning. The rightward shift estimate for chronic hypertension is a population approximation; individual patients may shift more or less than*

$$0.55 \times \Delta MAP_{chronic}.$$

11.8 The Software Module

Module 11.1 — Autoregulation Curve Simulator and Safe Pressure Range Calculator

Input: *Organ (brain / kidney / coronary); baseline chronic MAP; presence of acute event (stroke, hypertensive emergency); current MAP; myogenic gain k_m (default organ-specific values); maximum/minimum radius ratios.*

Processing:

1. *Compute $r_\infty(P)$ over pressure range 0–250 mmHg (equation 11.1).*

2. *Compute $Q(P)$ via Poiseuille with dynamic radius (equation 11.3).*

3. *Identify LLA and ULA from plateau region.*

4. *Compute rightward shift for chronic hypertension.*

5. *Calculate autoregulation index AI (equation 11.5).*

6. *For acute events: simulate impaired AI; compute safe MAP reduction range.*

7. *For ICP: compute CPP and safe MAP target.*

Output: *Autoregulation curve plot; LLA and ULA (normal and shifted); AI; safe MAP range for clinical scenario; CPP if ICP given.*

Full implementation at themathematicsoftheliving-body.com/modules.

11.9 Chapter Summary

1. *Autoregulation is local flow control: arterioles adjust resistance to maintain organ flow nearly constant over a wide pressure range. The autoregulation curve has an LLA, a plateau, and a ULA.*

2. *The myogenic response ODE: $\tau_m\, \mathrm{d}r/\,\mathrm{d}t = r_\infty(P) - r$ with $r_\infty(P) = r_0/[1 + k_m(P - P_0)]$. The autoregulation curve follows from Poiseuille's law with this dynamic radius.*

3. *The plateau condition ($\mathrm{d}Q/\,\mathrm{d}P = 0$) determines the autoregulation range. LLA and ULA are set by the minimum and maximum arteriolar radius capacities.*

4. *Chronic hypertension shifts the autoregulation curve rightward. The LLA rises from 60 to ≈ 80 mmHg. Rapid normalisation of MAP in a chronic hypertensive can cause cerebral ischaemia.*

5. *CPP = MAP $-$ ICP. The Cushing reflex is the baroreflex driving MAP to its maximum to maintain CPP above the LLA when ICP is critically elevated.*

6. *In the stroke worked example: estimated shifted LLA ≈ 80 mmHg; safe MAP target after acute stroke is no more than 15% reduction in 24 hours, targeting ≈ 140 mmHg for this patient.*

Key Equations, Chapter 11

$$r_\infty(P) = \frac{r_0}{1 + k_m(P - P_0)} \quad \textit{(myogenic response)}$$

$$Q(P) = \frac{\pi \Delta P\, r_0^4}{8\eta L\, [1 + k_m(P - P_0)]^4} \quad \textit{(autoregulation curve)}$$

$$AI = 1 - \frac{dQ/Q}{dP/P} \quad \textit{(autoregulation index)}$$

$$CPP = MAP - ICP$$

Key References, Chapter 11

1. *Lassen, N.A. (1959). Cerebral blood flow and oxygen consumption in man. Physiol. Rev. 39, 183–238. [Original description of cerebral autoregulation curve.]*

2. *Drummond, J.C. (1997). The lower limit of autoregulation: time to revise our thinking? Anesthesiology 86, 1431–1433. [Clinical LLA values and their variability.]*

3. *Czosnyka, M. et al. (1997). Continuous assessment of the cerebral vasomotor reactivity in head injury. Neurosurgery 41, 11–19. [Pressure reactivity index for clinical autoregulation monitoring.]*

Chapter 12

Exercise: The Cardiovascular System Under Load

Rest is the default state. Exercise is the stress test. Every component of the cardiovascular system — pump, pipes, regulators — reveals its true capacity only when pushed to its limits. The mathematics of exercise physiology is the mathematics of maximum performance.

— Zachariah Sinkala

12.1 The Physiology: Exercise as Integrative Stress Test

Exercise is the most powerful physiological challenge the cardiovascular system faces. In a sedentary adult at rest, cardiac output is ≈ 5 L/min. During maximal exercise in a trained athlete, it reaches ≈ 25 L/min — a *fivefold increase. Heart rate rises from 60 to 180 bpm. Stroke volume rises from 70 to $\approx$ 140 mL. Skeletal muscle blood flow rises fiftyfold. Oxygen consumption increases tenfold. All of this happens in seconds to minutes, coordinated by the same mathematical structures developed in Chapters 1–11.*

Exercise physiology is important clinically for three reasons:

- **The exercise stress test** *is the most commonly performed cardiovascular diagnostic test. Interpreting it requires understanding the expected cardiovascular response to graded exercise.*

- **Exercise capacity** *measured as $\dot{V}O_{2\,\max}$ is the strongest predictor of cardiovascular mortality, stronger than ejection fraction, blood pressure, or smoking status.*

- **Exercise training** *produces cardiovascular adaptations — the athletic heart — that represent the maximum physiological upregulation of the system.*

Before the equations: what happens in the skeletal muscle during exercise.

Exercise is initiated at the motor cortex, but the cardiovascular response begins even before the first muscle contraction — a phenomenon called central command. *Cortical motor neurons directly activate cardiovascular brainstem centres (RVLM, Chapter 9) in parallel with the motor signal, producing a feed-forward HR increase within the first 1–2 heartbeats of exercise.*

The skeletal muscle metabolic cascade. *Within contracting muscle fibres, ATP consumption rises immediately. Three metabolic pathways activate in sequence by energy yield: (1) Phosphocreatine hydrolysis ($PCr + ADP \rightarrow Cr + ATP$): within seconds, no O_2 required, limited by PCr stores (5–8 seconds at maximum intensity); (2) Anaerobic glycolysis: within 30 seconds, producing lactate; (3) Oxidative phosphorylation: dominant after 60–90 seconds, requires O_2 delivery.*

Metabolic vasodilation — the local signal. *As ATP is hydrolysed, metabolic byproducts accumulate: CO_2 ($\to H^+ \to$ vasodilation), K^+ (hyperpolarises VSMC via K_{ir}), adenosine (A_2A receptor on VSMC), and lactate (activates GPCR132/GPR81). These signals act in milliseconds — long before the sympathetic nervous system can respond — to dilate arterioles in active muscle 5–25-fold. This is why TPR falls so dramatically during exercise (from 22 to 4.8 Wood units, Chapter 12 table) before the sympathetic system fully activates.*

The nitric oxide-exercise link. *Shear stress on active muscle capillary endothelium rises dramatically with increased flow, activating eNOS (Chapter 14) and producing NO-mediated vasodilation. Regular exercise training chronically increases eNOS expression in muscle endothelium — one molecular mechanism underlying the reduced resting TPR and improved arterial compliance seen in trained individuals.*

12.2 The Observation: What Changes During Exercise

As exercise intensity increases from rest to maximum, the following changes occur in an orderly, mathematically describable sequence:

Variable	Rest	Mod.	Max.
HR (bpm)	60	130	195
SV (mL)	70	110	140
CO (L/min)	4.2	14.3	27.3
MAP (mmHg)	93	108	130
TPR (units)	22.1	7.6	4.8
SvO_2 (%)	75	50	20
$\dot{V}O_2$ (L/min)	0.25	2.0	5.0

Three observations stand out:

1. TPR falls dramatically. *Despite MAP rising, TPR falls from 22 to 4.8 Wood units — a 78% reduction. This is muscular vasodilation overwhelming the systemic vasoconstriction of the baroreflex. MAP rises despite falling TPR only because CO increases so dramatically.*

2. SvO_2 falls to 20%. *Tissues extract $\approx$ 80% of delivered oxygen — far above the resting 25%. This extreme extraction is the metabolic emergency signal that drives cardiovascular adaptation.*

3. CO rises steeply before autonomic activation. *In the first 10–15 seconds of exercise, before the sympathetic nervous system has fully activated, CO rises via the Frank-Starling mechanism (increased venous return from the muscle pump) and feed-forward central command (cortical activation of the cardiovascular centres).*

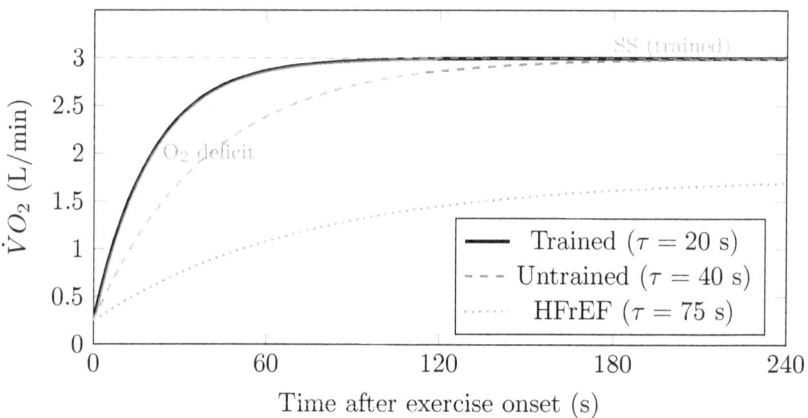

Figure 12.1: Oxygen uptake ($\dot{V}O_2$) kinetics after a step increase in exercise intensity. Trained athletes reach steady state rapidly ($\tau \approx 20$ s) because their cardiovascular system increases CO faster. Untrained individuals are slower ($\tau \approx 40$ s). Heart failure patients have severely impaired kinetics ($\tau > 60$ s) reflecting their limited CO reserve. The shaded area is the O_2 deficit, repaid via phosphocreatine hydrolysis and anaerobic glycolysis at exercise onset.

12.3 The Mathematics: The Exercise ODE System

12.3.1 The Cardiovascular Exercise ODE

Exercise physiology is a non-equilibrium steady state. Unlike rest, the cardiovascular system during exercise is continuously driven by a metabolic load signal $\dot{W}$ (mechanical power output, in watts). The governing ODE system integrates all components from Chapters 1–11:

$$\frac{\mathrm{d}\,CO}{\mathrm{d}t} = \frac{1}{\tau_{CO}}\left[HR(t) \cdot SV(t) - CO\right] \tag{12.1}$$

$$\frac{\mathrm{d}\,HR}{\mathrm{d}t} = \frac{1}{\tau_{HR}}\left[HR_\infty(\dot{W}, t) - HR\right] \tag{12.2}$$

$$\frac{\mathrm{d}\,TPR}{\mathrm{d}t} = \frac{1}{\tau_{TPR}}\left[TPR_\infty(\dot{W}) - TPR\right] \tag{12.3}$$

$$MAP(t) = CO(t) \times TPR(t) \tag{12.4}$$

where $\tau_{CO} \approx 15$ s, $\tau_{HR} \approx 30$ s, and $\tau_{TPR} \approx 60$ s are the response time constants of each variable, and the steady-state targets depend on workload $\dot{W}$.

12.3.2 Heart Rate and the Karvonen Formula

At steady-state exercise, heart rate is a linear function of workload (within the aerobic range):

$$HR_\infty(\dot{W}) = HR_{rest} + \frac{\dot{W}}{\dot{W}_{\max}} \cdot (HR_{\max} - HR_{rest}) \tag{12.5}$$

Maximum heart rate is estimated by:

$$HR_{\max} = 220 - age \text{ (years)} \tag{12.6}$$

The Karvonen formula for target HR at a given fraction f of maximum *aerobic capacity:*

$$HR_{target} = HR_{rest} + f \cdot (HR_{\max} - HR_{rest}) \tag{12.7}$$

At 70% intensity ($f = 0.70$) in a 40-year-old with $HR_{rest} = 65$ bpm:

$$HR_{target} = 65 + 0.70 \times (180 - 65) = 65 + 80.5 \approx 146 \; bpm$$

12.3.3 Oxygen Uptake Kinetics: The First-Order Model

During a step increase in exercise intensity, $\dot{V}O_2$ does not jump immediately to its new steady state. It follows first-order kinetics:

$$\dot{V}O_2(t) = \dot{V}O_{2,baseline} + \Delta\dot{V}O_2 \left(1 - e^{-t/\tau_{O_2}}\right) \tag{12.8}$$

where $\tau_{O_2} \approx 30\text{–}45$ s (in healthy adults) is the oxygen uptake time constant and $\Delta\dot{V}O_2$ is the steady-state increment.

A shorter τ_{O_2} indicates better cardiovascular fitness: the system reaches its new $\dot{V}O_2$ faster. In trained athletes, $\tau_{O_2} \approx 20$ s. In heart failure patients, $\tau_{O_2} > 60$ s — the slow oxygen kinetics reflect the failing heart's inability to rapidly increase CO in response to demand.

12.4 The Derivation: $\dot{V}O_{2\,\max}$ from the Fick Equation

$\dot{V}O_{2\,\max}$ is the maximum rate of oxygen consumption during exhaustive exercise — the gold standard measure of cardiovascular fitness. From the Fick equation (Chapter 6):

$$\dot{V}O_{2\,\max} = CO_{\max} \times (C_a - C_v)_{\max} \tag{12.9}$$

At maximum exercise:

- $CO_{\max} = HR_{\max} \times SV_{\max}$

- $(C_a - C_v)_{\max} \approx 16 \ mL \ O_2/dL$ *(tissues extract to $SvO_2 \approx 20\%$)*

Therefore:

$$\dot{V}O_{2\,\max} \ = \ HR_{\max} \times SV_{\max} \times 16 \times 10^{-1} \tag{12.10}$$

The three determinants of $\dot{V}O_{2\,\max}$ are $HR_{\max}$ (genetic, age-determined), $SV_{\max}$ (trainable), and $(C_a - C_v)_{\max}$ (determined by muscle oxidative capacity and haemoglobin).

For a healthy 35-year-old man:

$$\dot{V}O_{2\,\max} = 185 \times 140 \times 16 \times 0.1 \approx 4.1 \ L/min$$

$\dot{V}O_{2\,\max}$ per kg body weight (the standard clinical measure): $4.1/75 \approx 55 \ mL/kg/min$.

12.4.1   The Metabolic Equivalents (METs) Scale

One MET is the resting oxygen consumption: $\dot{V}O_{2,rest} = 3.5 \ mL/kg/min$. Exercise intensity in METs:

$$METs \ = \ \frac{\dot{V}O_2}{3.5 \ mL/kg/min} \tag{12.11}$$

Fitness classification by $\dot{V}O_{2\,\max}$: $\dot{V}O_{2\,\max} = 10 \ METs$ means the person can sustain $10 \times 3.5 = 35 \ mL/kg/min$. Patients unable to achieve 4 METs on exercise testing have poor surgical prognosis; $\geq 7 \ METs$ is associated with good cardiac prognosis.

12.5   The Clinical Interpretation: Reading the Exercise Test

12.5.1   The Exercise ECG

During a standard Bruce protocol treadmill test, 12-lead ECG is monitored continuously. The key diagnostic finding is ST-segment depression

≥ 1 *mm in two adjacent leads, indicating subendocardial ischaemia.*

Why does exercise provoke ischaemia? From Chapter 2, coronary flow depends on coronary perfusion pressure and coronary resistance. During exercise:

- *Heart rate increases $\rightarrow$ diastolic filling time shortens $\rightarrow$ less time for coronary perfusion (coronaries fill in diastole)*

- *Myocardial oxygen demand increases (HR $\times$ systolic pressure $=$ rate pressure product, the index of myocardial work)*

- *Coronary flow must increase 4–5-fold to meet demand*

A coronary stenosis that limits peak flow is unmasked during exercise. Below the ischaemic threshold, exercise-induced ST depression appears — the exercise test is positive.

12.5.2   The Rate-Pressure Product

The rate-pressure product (RPP) is the most practical index of myocardial oxygen demand:

$$RPP \;=\; HR \times P_{systolic} \quad [bpm \times mmHg] \tag{12.12}$$

Normal peak RPP during maximum exercise: $\approx 180 \times 200 = 36{,}000$. ST depression typically appears at a reproducible RPP threshold for a given patient. If ST depression occurs at $RPP = 20{,}000$ in one test and $RPP = 20{,}500$ in the next (after an intervention), the ischaemic threshold has not meaningfully changed.

The RPP threshold at which ischaemia appears is inversely related to the severity of coronary disease — the more severe the stenosis, the lower the threshold.

12.5.3 $\dot{V}O_{2\,\mathrm{max}}$ as a Prognostic Marker

$\dot{V}O_{2\,\mathrm{max}}$ predicts mortality independent of all other risk factors. From the Fick equation, low $\dot{V}O_{2\,\mathrm{max}}$ reflects either:

1. *Low CO_{max} (cardiac limitation: impaired HR, SV, or both)*

2. *Low $(C_a - C_v)_{max}$ (peripheral limitation: reduced muscle oxidative capacity, anaemia)*

Distinguishing cardiac from peripheral limitation requires measuring CO during exercise (by thermodilution or acetylene rebreathing). If CO increases normally but $\dot{V}O_{2\,max}$ is low, the limitation is peripheral. If CO reaches its maximum at a lower workload than expected, the limitation is cardiac.

Why $\dot{V}O_{2\,max}$ predicts mortality better than ejection fraction: *EF measures resting pump function. $\dot{V}O_{2\,max}$ measures the entire cardiovascular chain under maximum stress — pump, vasculature, autoregulation, baroreflex, oxygen transport, and mitochondrial capacity simultaneously. It is the integral of everything developed in Chapters 1–11. A person with preserved EF but severely impaired cardiac output reserve (unable to increase CO during exercise) will have a low $\dot{V}O_{2\,max}$ and a poor prognosis — which EF alone would never have revealed.*

12.6 The Athletic Heart: Training Adaptations

Regular endurance exercise produces structural and functional cardiovascular adaptations that maximise $\dot{V}O_{2\,max}$:

Cardiac adaptations:

- *Increased LV end-diastolic volume (eccentric hypertrophy, increased preload capacity): EDV rises by $\approx 20\text{–}30\%$*

- *Increased SV at rest and during exercise (from Frank-Starling, Chapter 5)*

- *Resting bradycardia (HR = 40–50 bpm in elite athletes): increased vagal tone and reduced intrinsic SAN rate*

Vascular adaptations:

- *Increased capillary density in skeletal muscle (more surface area for O_2 delivery)*

- *Increased arterial compliance (lower pulse pressure for same SV, from Chapter 3)*

- *Enhanced flow-mediated vasodilation (improved endothelial function)*

The mathematical result: *From the Fick equation (12.9), each adaptation has a specific contribution:*

$$\frac{\Delta \dot{V}O_{2\,\mathrm{max}}}{\dot{V}O_{2\,\mathrm{max}}} = \frac{\Delta SV_{\mathrm{max}}}{SV_{\mathrm{max}}} + \frac{\Delta(C_a - C_v)_{\mathrm{max}}}{(C_a - C_v)_{\mathrm{max}}} \qquad (12.13)$$

(HR_{max} does not change with training.) A 20% increase in SV_{max} and a 10% increase in $(C_a - C_v)_{\mathrm{max}}$ produce a 30% increase in $\dot{V}O_{2\,\mathrm{max}}$ — consistent with the observed training response in previously sedentary adults.

12.7 The Worked Example: Interpreting an Exercise Stress Test

A 58-year-old woman with hypertension and type 2 diabetes undergoes an exercise stress test for evaluation of atypical chest pain. Bruce protocol. HR_{max} predicted: $220 - 58 = 162$ bpm.

Stage 1 (1.7 mph, 10% grade, 5 METs): *HR 112 bpm, MAP 148 mmHg, no symptoms, no ECG changes. $RPP = 112 \times 188 = 21{,}056$.*

Stage 2 (2.5 mph, 12% grade, 7 METs): *HR 138 bpm, MAP 162 mmHg, chest tightness, 1.5 mm horizontal ST depression in V4–V6. $RPP = 138 \times 204 = 28{,}152$.*

Test stopped. Peak $HR = 138$ bpm $= 85\%$ of HR_{max} (adequate effort).

Step 1 — Classify the test. *ST depression ≥ 1 mm in ≥ 2 leads with symptoms at $< 85\%$ HR_{max}: **positive for inducible ischaemia**. Duke Treadmill Score:*

$$DTS \; = \; Exercise\ time - 5 \times ST\ deviation - 4 \times Angina\ index \quad (12.14)$$

$$DTS = 6 - 5 \times 1.5 - 4 \times 1 = 6 - 7.5 - 4 = -5.5$$

DTS < -11: *high risk. DTS* -10 *to* $+4$: *intermediate risk. DTS* $> +5$: *low risk. DTS* $= -5.5$: **intermediate risk**.

Step 2 — Estimate functional capacity. *Stopped at Stage 2: estimated* $\dot{V}O_{2\,max} \approx 7$ *METs* $= 24.5$ *mL/kg/min. For her age/sex, the expected median is 28 mL/kg/min. She is at the 30th percentile for fitness — below average, consistent with diabetes and hypertension.*

Step 3 — Ischaemic threshold. *RPP at ischaemia onset:* $\approx 28{,}000$. *This is the patient-specific threshold for coronary supply-demand mismatch — reproducible on repeat testing and used to assess response to medical therapy or revascularisation.*

Step 4 — Management. *Intermediate DTS: refer for coronary CT angiography or nuclear stress imaging to quantify ischaemic burden before deciding on invasive evaluation.*

> **Model Assumptions, Chapter 12.** *What the exercise model assumes: (1) Oxygen uptake kinetics follow a single exponential with time constant* τ_{O_2}. *(2) Heart rate is a linear function of workload within the aerobic range. (3)* $\dot{V}O_{2\,max}$ *is limited by cardiac output (not peripheral utilisation). Where it breaks down: Real* $\dot{V}O_2$ *kinetics are better described by two exponentials (fast and slow components). HR-workload linearity breaks down above the anaerobic threshold. Many trained athletes are peripherally limited —* $\dot{V}O_{2\,max}$ *is limited by muscle mitochondrial capacity, not CO. The* $HR_{max} = 220 - age$ *formula has* ± 10 *bpm SD at any given age.*

12.8 The Software Module

Module 12.1 — Exercise Cardiovascular Simulator and Stress Test Interpreter

Input: Age, sex, weight, height; resting HR and MAP; exercise protocol (Bruce stages or custom watts); ECG findings at each stage (ST deviation in mm, symptoms); peak HR achieved; optional: CO measurement during exercise.

Processing:

1. *Compute $HR_{\max}$, Karvonen target HRs (equations 12.6, 12.7).*

2. *Solve exercise ODE system (equations 12.1–12.3) for each stage.*

3. *Compute $\dot{V}O_{2\,\max}$ estimate from peak stage achieved; classify fitness percentile for age/sex.*

4. *Compute RPP at each stage; identify ischaemic threshold if ST changes present.*

5. *Compute Duke Treadmill Score (equation 12.14); classify risk.*

6. *If CO measured: compute Fick-derived $\dot{V}O_{2\,\max}$; distinguish cardiac from peripheral limitation.*

Output: Exercise haemodynamics time course; $\dot{V}O_{2\,\max}$ and fitness percentile; RPP time course and ischaemic threshold; Duke score and risk classification; cardiac vs. peripheral limitation assessment.

Full implementation at themathematicsoftheliving-body.com/modules.

12.9 Chapter Summary

1. *Exercise increases CO from 5 to 25 L/min (fivefold), HR from 60 to 195 bpm, and decreases TPR by 78%. MAP rises moderately because CO increase exceeds TPR fall.*

2. *The exercise ODE system integrates Frank-Starling (SV), autonomic activation (HR), and metabolic vasodilation (TPR) into a coupled first-order system with time constants of 15–60 s per variable.*

3. $\dot{V}O_{2\,\max} = CO_{\max} \times (C_a - C_v)_{\max}$ *(Fick). Its three determinants are* $HR_{\max}$ *(genetic),* $SV_{\max}$ *(trainable), and peripheral* O_2 *extraction (trainable).*

4. *Ischaemic threshold on exercise testing is reproducible for a given patient and defined by the RPP at which ST depression appears. Duke Treadmill Score quantifies overall risk.*

5. $\dot{V}O_{2\,\max}$ *predicts mortality better than EF because it tests the entire cardiovascular chain under maximum load — integrating every component from Chapters 1–11.*

6. *Endurance training increases* $\dot{V}O_{2\,\max}$ *by* $\approx 30\%$ *through increased* $SV_{\max}$ *(eccentric hypertrophy) and peripheral extraction, with no change in* $HR_{\max}$.

Key Equations, Chapter 12

$$\dot{V}O_{2\,\max} = CO_{\max} \times (C_a - C_v)_{\max} \quad (Fick)$$

$$HR_{target} = HR_{rest} + f(HR_{\max} - HR_{rest}) \quad (Karvonen)$$

$$RPP = HR \times P_{systolic} \quad (myocardial\ demand)$$

$$DTS = t_{ex} - 5\,\Delta ST - 4\,angina\ index$$

Key References, Chapter 12

1. *Fick, A. (1870). (as Chapter 6 —* $\dot{V}O_{2\,\max}$ *uses the same Fick principle.)*

2. *Gibbons, R.J. et al. (1997). ACC/AHA Guidelines for Exercise Testing. J. Am. Coll. Cardiol. 30, 260–311. [Clinical exercise testing standards, Duke Treadmill Score.]*

3. Myers, J. et al. (2002). *Exercise capacity and mortality among men referred for exercise testing. N. Engl. J. Med. 346, 793–801.* [$\dot{V}O_{2\,\mathrm{max}}$ *as the strongest predictor of cardiovascular mortality.*]

Chapter 13

Heart Failure as a Dynamical Disease

Heart failure is not the heart stopping. It is the heart stuck in a new equilibrium — one where every compensatory mechanism that was designed to restore function has become the source of further dysfunction. Understanding this requires dynamics, not statics.

— *Zachariah Sinkala*

13.1   The Physiology: Opening Part IV

Parts I–III built the normal cardiovascular system: its physics, mechanics, electrical behaviour, and regulatory architecture. Part IV examines what happens when this system fails — four canonical disease states that represent parameter shifts, setpoint changes, or loss of stability in the dynamical models already developed.

Heart failure is the first and most important. It is not a single disease but a syndrome: any condition that impairs the heart's ability to provide cardiac output sufficient for the body's metabolic needs. It affects ≈ 6.5 million Americans and is the leading cause of hospitalisation in adults over 65.

The medical student learns heart failure as a list of symptoms (dyspnoea, oedema, fatigue) and a list of treatments (diuretics, ACE inhibitors, beta-blockers, devices). This chapter provides what the list cannot: why each symptom arises from a specific parameter change in the models of Chapters 4–11, and why each treatment works by correcting a specific parameter.

Before the equations: what fails at the cellular level in heart failure.

Heart failure is not simply a weak pump. At the cardiomyocyte level, multiple parallel defects accumulate:

Excitation-contraction coupling impairment. *In HFrEF, the L-type Ca^{2+} channel (Cav1.2) density decreases and its phosphorylation by CaMKII and PKA is dysregulated. Ryanodine receptor 2 (RyR2) is hyperphosphorylated by CaMKII, causing spontaneous SR Ca^{2+} leak during diastole (the substrate for DAD-mediated arrhythmias, Chapter 8) and depleting SR Ca^{2+} stores. SERCA2a expression decreases and phospholamban inhibition of SERCA2a increases, slowing Ca^{2+} reuptake (impaired lusitropy — the failing heart relaxes slowly as well as contracts weakly).*

Myofilament protein changes. *α-Myosin heavy chain (MYH6, the fast, high ATPase isoform) is downregulated and β-myosin heavy chain (MYH7, the slow, energy-efficient isoform) is upregulated in the failing ventricle. This isoform switch reduces E_{es} directly — β-MHC has lower cross-bridge cycling velocity and generates less force per ATP. Titin is also spliced toward stiffer isoforms, increasing passive stiffness (contributing to HFpEF phenotype).*

Energy substrate shift. *The normal heart derives $\approx 70\%$ of its*

ATP from fatty acid oxidation and $\approx 30\%$ from glucose. The failing heart shifts to almost exclusively glucose metabolism — partially because PPARα (the transcription factor driving fatty acid oxidation genes) is downregulated. Glucose metabolism is less oxygen-efficient (5.6 ATP/O_2 vs 5.9 for fatty acids), further impairing mechanical efficiency ($\eta = W_{stroke}/PVA$, Chapter 4).

13.2 The Observation: Two Phenotypes

Heart failure presents in two distinct phenotypes, each with a different primary abnormality:

***HFrEF (Heart Failure with reduced Ejection Fraction, EF <** 40%**):** The ventricle cannot contract adequately. E_{es} (Chapter 4) is reduced. The ESPVR shifts rightward and flattens. ESV is high, SV is low.*

***HFpEF (Heart Failure with preserved Ejection Fraction, EF** $\geq 50\%$**):** The ventricle cannot relax adequately. Diastolic stiffness (Chapter 4 EDPVR) is increased. EDV is normal but filling pressures are elevated. SV may be normal at rest but cannot increase during exercise.*

Figure 13.1 shows how each phenotype deforms the normal PV loop.

13.3 The Mathematics: Heart Failure as a Parameter Shift

13.3.1 HFrEF: The ESPVR Shifts

From Chapter 4, stroke volume is: $SV = EDV - V_d - P_{AO}/E_{es}$. In HFrEF, E_{es} falls from ≈ 2.0 to ≈ 0.4–0.8 mmHg/mL. The consequences:

1. ESV rises: from 50 mL (normal) to ≈ 185 mL

2. SV falls: from 70 mL to ≈ 55 mL

3. EF falls: from 58% to 23%

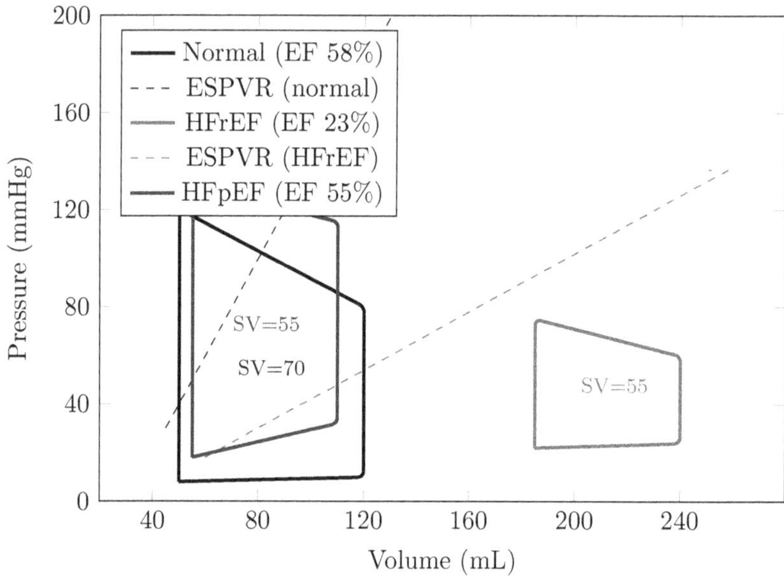

Figure 13.1: Pressure-volume loops for three conditions. **Normal** (black): EDV 120 mL, ESV 50 mL, SV 70 mL, EF 58%. **HFrEF** (red): dilated ventricle, reduced E_{es} (ESPVR flattened and shifted), EDV 240 mL, SV 55 mL, EF 23%. **HFpEF** (blue): normal size but stiff diastole (EDPVR steeper), elevated filling pressures, EF 55% but SV falls with exercise.

4. CO falls: from 4.2 to $\approx$ 3.1 L/min (HR compensates partially)

The ventricle dilates chronically (EDV rises from 120 to 240 mL) — a Frank-Starling compensation that partially restores SV via the ascending limb. But the compensation is incomplete and costly: the dilated ventricle has increased wall stress (Laplace, Chapter 5), higher myocardial oxygen demand, and eventually further depresses E_{es}.

13.3.2 Compensation and Decompensation: A Dynamical Transition

The progression of HFrEF from compensated to decompensated is a dynamical transition — a shift from one stable equilibrium to another. The compensatory mechanisms are:

Mechanism	*Benefit*	*Long-term cost*
Frank-Starling	$\uparrow SV$	*Wall stress, O_2 demand*
Tachycardia	$\uparrow CO$	$\downarrow$ *diastolic fill*
RAAS activation	$\uparrow volume$	*Pulmonary oedema*
SNS activation	$\uparrow inotropy$	*Arrhythmia, toxicity*
Hypertrophy	$\uparrow E_{es}$	$\downarrow$ *compliance*

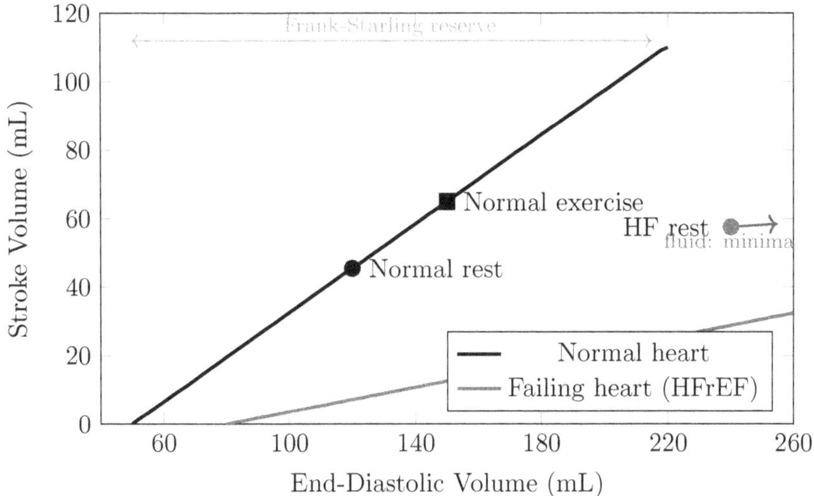

Figure 13.2: Frank-Starling curves for the normal and failing heart. The normal heart (black) has a steep ascending limb: increasing EDV substantially increases SV. The failing heart (red) has a flat curve: compensatory dilation (high EDV) yields minimal additional SV. A fluid challenge in HFrEF produces little gain while elevating filling pressures — explaining why volume management in heart failure requires careful titration.

13.4 The Derivation: Why Compensation Fails

13.4.1 The Neurohormonal Vicious Cycle

Reduced CO activates the RAAS (Chapter 10) and the sympathetic nervous system simultaneously. Both initially compensate but ultimately accelerate deterioration. Figure 13.3 maps this cycle mathematically.

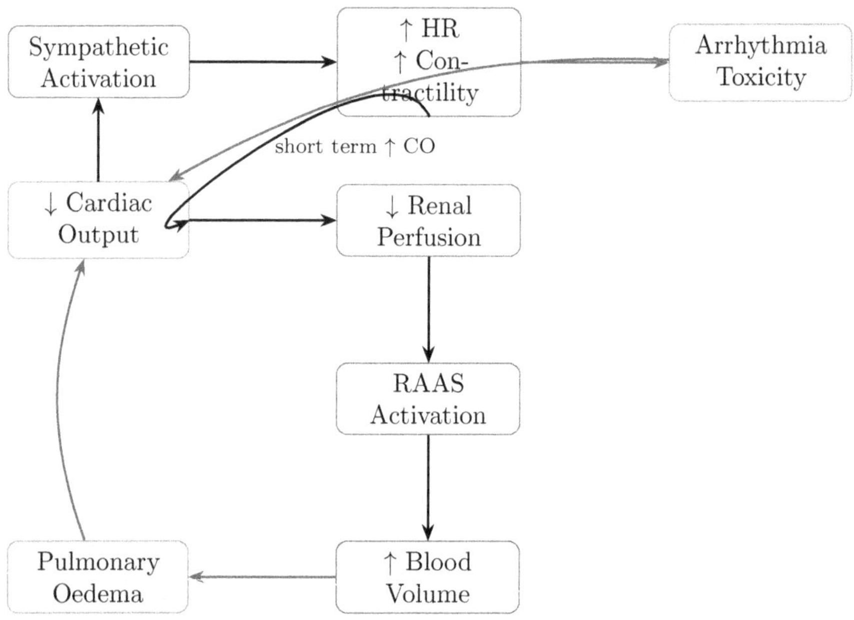

Figure 13.3: Neurohormonal vicious cycle in HFrEF. Reduced CO activates RAAS (volume overload) and SNS (tachycardia). Short-term compensation leads to long-term arrhythmia and myocyte toxicity.

The dynamics of this cycle are described by the coupled ODE system combining Chapters 4, 5, 9, and 10:

$$\frac{\mathrm{d}\,CO}{\mathrm{d}t} = \frac{HR \cdot SV(E_{es}, EDV, P_{AO}) - CO}{\tau_c} \qquad (13.1)$$

$$\frac{\mathrm{d}\,EDV}{\mathrm{d}t} = \frac{V_b/V_{ref} - EDV}{\tau_v} \qquad (13.2)$$

$$\frac{\mathrm{d}\,V_b}{\mathrm{d}t} = k_{Na}[Aldo] - k_{loss}V_b \qquad (13.3)$$

*where V_b is blood volume (from the RAAS ODE, Chapter 10) and $\tau_v \approx$
2 hours is the time constant for ventricular filling to respond to volume
changes.*

*The system has two stable equilibria: the compensated state (where Frank-
Starling, RAAS, and SNS activation maintain near-normal CO at the cost
of elevated filling pressures) and the decompensated state (where filling
pressures rise above pulmonary capillary wedge pressure 18 mmHg and
pulmonary oedema occurs).*

*The transition between these states is triggered by acute events: myocar-
dial ischaemia, arrhythmia, non-compliance with medications, or dietary
sodium excess — each representing a perturbation that pushes the system across the basin of attraction boundary from the compensated to the
decompensated equilibrium.*

13.5 The Clinical Interpretation: Matching Treatment to Parameter

*Every evidence-based heart failure therapy corrects a specific parameter in
the dynamical model:*

Drug / Device	Parameter	Mechanism
ACEi / ARB	$\downarrow P_{AO}$	Reduces afterload, cuts RAAS
Beta-blocker	$\uparrow E_{es}$	Reverses SNS toxicity, $\uparrow \beta_1$
MRA (spiro)	$\downarrow[Aldo]$	Anti-fibrotic, reduces volume
SGLT2i	$\downarrow V_b$	Osmotic diuresis, reduces preload
Loop diuretic	$\downarrow EDV$	Acute volume reduction, relieves oedema
ICD	Arrhythmia	Prevents VF as terminal event
CRT	$\uparrow E_{es}$	Resynchronises contraction

Why beta-blockers — a drug that slows and weakens the heart — improve survival in heart failure: *In acute decompensation, giving a beta-blocker seems counterintuitive: it reduces HR and E_{es}, worsening CO acutely. But in chronic HFrEF, the SNS is chronically over-activated. Chronically elevated catecholamines downregulate β_1 receptors, cause myocyte toxicity, and promote arrhythmias. Beta-blockers block this toxicity. Over 3–6 months, E_{es} actually increases (receptor upregulation, reverse remodelling), HR falls, and EF improves. The counterintuitive result follows directly from the ODE: short-term $\Delta E_{es} < 0$, long-term $\Delta E_{es} > 0$ as the SNS toxicity is reversed.*

13.6 The Worked Example: Quantifying Acute Decompensation

A 67-year-old man with known HFrEF (EF 32%, $E_{es} = 0.7$ mmHg/mL) presents with acute pulmonary oedema. Pulmonary artery catheter: wedge pressure (PCWP) 32 mmHg, CO 2.8 L/min, MAP 78 mmHg, HR 118 bpm.

Step 1 — Characterise the haemodynamic state. *Thermodilution CO has ±15% measurement variability; the true CO range is 2.38– 3.22 L/min.*

$$CI = \frac{2.8}{1.85} = 1.51 \ L/min/m^2 \quad (range\ 1.29\text{–}1.74\ from\ CO\ uncertainty) \quad (critical:$$

$$TPR = \frac{78}{2.8/60} = \frac{78}{0.047} = 1{,}660 \ dyn \cdot s/cm^5 \quad (elevated)$$

This is the classic Forrester Class IV haemodynamic profile: low CO, high filling pressure ("wet and cold").

Step 2 — Predict SV from the PV loop model.

$$SV = EDV - V_d - \frac{P_{AO}}{E_{es}} = 215 - 30 - \frac{78}{0.7} = 185 - 111 = 74 \ mL$$

$$HR_{estimated} = \frac{CO}{SV} = \frac{2{,}800}{74} \approx 38 \ bpm?$$

Discrepancy: the Fick-derived CO of 2.8 L/min with measured HR 118 implies $SV = 2800/118 = 24$ mL, far below the PV-loop predicted 74 mL. Resolution: acute mitral regurgitation is likely — some stroke volume is regurgitating into the left atrium rather than into the aorta. Forward SV is only ≈ 24 mL; total SV is 74 mL; regurgitant fraction $\approx 68\%$.

Step 3 — Treatment plan from the model.

> **Model limit:** *The single-compartment PV loop model (Chapter 4) assumes all stroke volume is ejected forward into the aorta. In mitral regurgitation, a fraction f_r of SV is ejected retrograde into the left*

> *atrium. The model must be corrected: $CO_{forward} = (1 - f_r) \times SV \times HR$. Without this correction, the model substantially overestimates forward CO. f_r must be estimated from echo (regurgitant fraction) or by comparing Fick CO to thermodilution CO.*

Targets: $PCWP < 18$ mmHg, $CI > 2.2$ L/min/m^2.

- *IV nitroprusside: reduces $P_{AO} \to$ increases total SV and reduces regurgitant fraction f_r (lower afterload makes the forward path preferential) $\to$ net increase in forward CO*

- *IV furosemide: reduces EDV, reduces PCWP*

- *Do NOT give beta-blocker acutely*

Using the corrected model with estimated $f_r = 0.68$ at baseline:

$$CO_{forward,\ baseline} = (1 - 0.68) \times 0.074 \times 118 = 0.32 \times 8.7 = 2.8 \ L/min \quad \checkmark$$

After nitroprusside ($P_{AO} : 78 \to 65$ mmHg, f_r falls to ≈ 0.40 as afterload is reduced):

$$SV_{new} = 215 - 30 - \frac{65}{0.7} = 92 \ mL$$

$$CO_{forward,\ new} = (1 - 0.40) \times 0.092 \times 118 \approx 6.5 \ L/min$$

This is plausible and clinically significant. The key model insight: in mitral regurgitation, afterload reduction simultaneously increases total SV and redirects it forward — a double benefit that the corrected model predicts.

> **Model Assumptions, Chapter 13.** *What the heart failure model assumes: (1) The PV loop ESPVR is linear with a single E_{es} value. (2) The Frank-Starling curve has a stable ascending limb (no right ventricular failure complicating LV filling). (3) The neurohormonal cascade can be approximated as a linear ODE system. Where it breaks down: In acute decompensation, E_{es} changes rapidly (from acidosis, ischaemia, stunning) in ways the single-parameter model cannot track. Bi-ventricular failure (right heart failure reducing LV*

preload) is not captured. The neurohormonal ODE is a teaching abstraction — real neurohumoral biology involves dozens of interacting mediators. All CO estimates from thermodilution carry $\pm 15\%$ uncertainty, which propagates to $\pm 25\%$ uncertainty in the CI estimate.

13.7 The Software Module

Module 13.1 — Heart Failure Haemodynamic Simulator

Input: *EF, E_{es}, EDV, ESV; PCWP; CO (by thermodilution or Fick); MAP, HR; drug: ACEi/ARB inhibition, beta-blocker dose, diuretic (target PCWP reduction).*

Processing:

1. *Classify Forrester haemodynamic profile (I–IV) from CO and PCWP.*

2. *Solve PV loop ODE for current state; compute SV, E_{es}, wall stress.*

3. *Integrate neurohormonal ODE system (equations 13.1–13.3) over 6–12 hours.*

4. *Apply treatment: modify P_{AO}, EDV, or E_{es} per drug; recompute CO and PCWP.*

5. *Plot PV loop and Frank-Starling curve before and after intervention.*

Output: *Forrester class; CO, CI, PCWP; PV loop comparison; Frank-Starling operating point; treatment response prediction; neurohormonal time course.*

Full implementation at themathematicsoftheliving-body.com/modules.

13.8 Chapter Summary

1. *Heart failure is a dynamical disease: two phenotypes (HFrEF, reduced E_{es}; HFpEF, increased diastolic stiffness) produce distinct PV loop deformations (Figure 13.1).*

2. *The Frank-Starling curve in HFrEF is flat: compensatory dilation produces minimal SV gain at the cost of elevated filling pressure (Figure 13.2).*

3. *The neurohormonal vicious cycle (Figure 13.3): RAAS and SNS activation compensate acutely but accelerate decompensation chronically through volume overload, arrhythmia, and myocyte toxicity.*

4. *Every evidence-based heart failure treatment targets a specific parameter in the dynamical model: ACEi/ARB ($\downarrow P_{AO}$), beta-blockers (reverse E_{es} decline), MRA ($\downarrow$[Aldo]), SGLT2i ($\downarrow V_b$), diuretics ($\downarrow$EDV).*

5. *Beta-blockers improve survival because chronic SNS over-activation depresses E_{es} through receptor downregulation and myocyte toxicity. Blocking it produces reverse remodelling and E_{es} recovery over months — counterintuitive from the acute pharmacodynamics, predictable from the ODE.*

6. *The worked example demonstrated acute decompensation diagnosis (Forrester Class IV, CI 1.51) and mathematical treatment planning: nitroprusside reduces P_{AO}, increases forward SV, and reduces regurgitant fraction — all predicted from the PV loop model.*

Key Equations, Chapter 13

$$SV = EDV - V_d - P_{AO}/E_{es}$$

$$\frac{\mathrm{d}\,EDV}{\mathrm{d}t} = (V_b/V_{ref} - EDV)/\tau_v$$

$$CPP = MAP - PCWP \quad \text{(coronary perfusion pressure)}$$

$$CI = CO/BSA \quad (< 1.8 = haemodynamic\ compromise)$$

Key References, Chapter 13

1. *Braunwald, E. & Bristow, M.R. (2000). Congestive heart failure: fifty years of progress. Circulation 102 (Suppl 4), IV-14–IV-23. [Historical overview of heart failure pathophysiology.]*

2. *Packer, M. et al. (2001). Effect of carvedilol on survival in severe chronic heart failure. N. Engl. J. Med. 344, 1651–1658. [COPERNICUS trial — β-blocker survival benefit in severe HFrEF.]*

3. *Forrester, J.S. et al. (1976). Medical therapy of acute myocardial infarction by application of hemodynamic subsets. N. Engl. J. Med. 295, 1356–1362. [Forrester haemodynamic classification used in worked example.]*

Chapter 14

Hypertension: A Setpoint Shift

The kidney does not malfunction in hypertension. It functions perfectly — at the wrong setpoint. Understanding why the setpoint shifts requires knowing which cells in the kidney are reading the pressure signal, what proteins they use to read it, and what happens to those proteins when they are chronically exposed to a world that evolution never prepared them for.

— Zachariah Sinkala

14.1 The Biology First: What Hypertension Actually Is at the Cellular Level

Before the equations: the cells that set blood pressure.

Blood pressure is dominated in the long run *by the kidney — specifically, by the juxtaglomerular apparatus (JGA), a cluster of specialised cells at the junction of the afferent arteriole and the distal tubule that continuously monitors renal perfusion pressure and sodium delivery.*

The JGA contains three cell types that collectively implement the pressure-natriuresis system:

***Juxtaglomerular (JG) cells** in the afferent arteriole wall are modified smooth muscle cells that contain renin granules. They sense wall stress (transmural pressure) via mechanosensitive channels — primarily the Piezo1 mechanoreceptor, a 2,500-amino-acid trimeric ion channel that opens when the membrane is stretched. When renal perfusion pressure falls, Piezo1 channel opening decreases, intracellular Ca^{2+} falls, and renin is released by exocytosis. When pressure rises, Piezo1 activity increases, intracellular Ca^{2+} rises, and renin secretion is suppressed.*

***Macula densa cells** in the thick ascending limb sense luminal NaCl via the NKCC2 cotransporter ($Na^+/K^+/2Cl^-$ cotransporter, encoded by SLC12A1). When NaCl delivery falls, NKCC2 transport decreases, intracellular $[Cl^-]$ falls, and a paracrine signal (prostaglandin E_2, nitric oxide) stimulates JG cell renin release. This is the tubuloglomerular feedback loop — a local circuit within the nephron.*

***Endothelial cells** of the afferent arteriole produce nitric oxide (NO) via eNOS (endothelial nitric oxide synthase, encoded by NOS3) in response to shear stress. NO diffuses into vascular smooth muscle, activates soluble guanylate cyclase, raises cGMP, and causes vasodilation.*

> ***Essential hypertension*** *develops when one or more of these setpoint-determining mechanisms shifts: Piezo1 sensitivity changes, NKCC2 expression increases, eNOS activity decreases, or the smooth muscle response to NO is attenuated. The result is that the kidney retains sodium and water at a higher pressure than it did before —— the pressure-natriuresis curve shifts rightward.*

14.2 The Observation: The Pressure-Natriuresis Curve

The pressure-natriuresis curve is the single most important concept in understanding long-term blood pressure control. It plots urinary sodium excretion (mEq/day) against mean arterial pressure (mmHg). At equilibrium, sodium intake must equal sodium excretion. The body's operating point is where the sodium intake line intersects the pressure-natriuresis curve.

Three physiological facts about this curve:

1. ***It has a steep slope.*** *A 10 mmHg rise in MAP approximately doubles urinary sodium excretion in the normal kidney. This steep slope is why the kidney is so effective at long-term pressure control.*

2. ***It shifts rightward in hypertension.*** *The hypertensive kidney excretes less sodium at any given pressure than a normal kidney. To achieve sodium balance, MAP must be higher. This rightward shift is the definition of hypertension from a renal physiology perspective.*

3. ***Antihypertensive drugs work by shifting it back leftward.*** *Every effective antihypertensive —— diuretic, ACE inhibitor, ARB, or calcium channel blocker —— shifts the pressure-natriuresis curve leftward, allowing sodium balance at a lower MAP.*

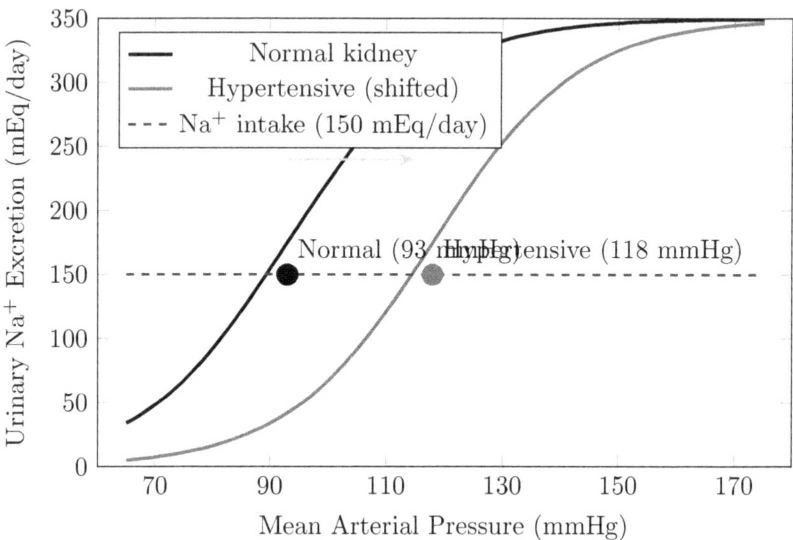

Figure 14.1: The pressure-natriuresis curve. At sodium intake of 150 mEq/day, the normal kidney achieves balance at MAP 93 mmHg (black dot). The hypertensive kidney's curve is shifted 25 mmHg rightward: balance is only achieved at MAP 118 mmHg. Treatment shifts the curve back leftward.

14.3 The Mathematics: Guyton's Model of Long-Term Pressure Control

14.3.1 The Pressure-Natriuresis Equation

The pressure-natriuresis relationship is modelled as:

$$\dot{U}_{Na}(P) = \dot{U}_{Na}^0 \cdot \left(\frac{P}{P_0}\right)^n \tag{14.1}$$

where $\dot{U}_{Na}$ is urinary sodium excretion (mEq/day), P_0 is the normal operating pressure (93 mmHg), $\dot{U}_{Na}^0$ is the normal excretion rate matching intake (150 mEq/day), and n is the slope exponent ($n \approx 3$–5 in normal subjects, lower in salt-sensitive hypertension).

At equilibrium, excretion equals intake I_{Na}:

$$I_{Na} = \dot{U}^0_{Na} \cdot \left(\frac{P^*}{P_0}\right)^n \tag{14.2}$$

Solving for the equilibrium pressure:

[Approximation — see Model Assumptions]

$$P^* = P_0 \cdot \left(\frac{I_{Na}}{\dot{U}^0_{Na}}\right)^{1/n} \tag{14.3}$$

For $n = 3$ and sodium intake doubled ($I_{Na} = 300$ mEq/day):

$$P^* = 93 \times (300/150)^{1/3} = 93 \times 2^{1/3} = 93 \times 1.26 \approx 117 \ mmHg$$

A doubling of sodium intake raises equilibrium MAP by only 26% in a normal kidney ($n = 3$). In salt-sensitive hypertension ($n = 1$), the same sodium doubling doubles the pressure: $P^* = 93 \times 2 = 186$ mmHg. This is the mathematical definition of salt sensitivity: a small exponent n in the pressure-natriuresis relationship.

14.3.2 The Molecular Basis of Salt Sensitivity

Why some people are salt-sensitive: the molecular story.

Salt sensitivity affects $\approx 50\%$ of hypertensives. It reflects reduced efficiency of the pressure-natriuresis response — a smaller n in equation (14.1). Three molecular mechanisms are established:

1. Reduced eNOS activity. *The NOS3 gene encodes eNOS. The Glu298Asp variant (rs1799983) reduces eNOS expression by $\approx 30\%$. Less NO in the renal vasculature means less pressure-induced vasodilation of the afferent arteriole, reducing the natriuretic response to pressure. Carriers have $n \approx 1.5$ versus $n \approx 3.5$ in non-carriers.*

2. ENaC overactivity. *The epithelial sodium channel (ENaC, encoded by SCNN1A/B/G) in the cortical collecting duct is the final sodium reabsorption step. Liddle syndrome (gain-of-function*

ENaC mutation) and mineralocorticoid excess (primary aldosteronism) cause ENaC overactivity — the kidney reabsorbs sodium regardless of pressure, reducing n toward zero (extreme salt sensitivity).

3. Impaired tubuloglomerular feedback. *NKCC2 variants and macula densa signalling defects reduce the sensitivity of tubuloglomerular feedback, blunting the natriuretic response to increased NaCl delivery.*

This molecular understanding has a direct clinical application: patients with reduced eNOS activity respond poorly to calcium channel blockers (which act on vascular smooth muscle, not eNOS) but respond well to ACE inhibitors (which increase bradykinin, which stimulates eNOS). Genetic profiling of NOS3 variants could predict drug response.

14.4   The Derivation: The Guyton Diagram

Guyton's long-term model of blood pressure integrates cardiac function and renal function into a single framework. The cardiac output curve (CO as a function of MAP) and the pressure-natriuresis curve interact to set both MAP and CO simultaneously.

At equilibrium, two conditions must be satisfied:

1. ***Cardiac:*** *CO is determined by venous return, which is determined by blood volume V_b and the venous circuit:*

$$CO = f_c(V_b, TPR) \tag{14.4}$$

2. ***Renal:*** *Blood volume is determined by sodium balance, which is determined by MAP:*

$$\frac{dV_b}{dt} = k_v \left[I_{Na} - \dot{U}_{Na}(P) \right] \tag{14.5}$$

At steady state ($dV_b/dt = 0$): $I_{Na} = \dot{U}_{Na}(P^*)$, giving P^* from equa-

tion (14.3).

The Guyton model predicts that long-term blood pressure is set entirely by the pressure-natriuresis relationship — not by TPR, not by cardiac function, not by the baroreflex. These other factors can shift MAP acutely but not chronically, because chronically the kidney compensates via volume.

This is Guyton's infinite gain *principle: the long-term renal control of blood pressure has infinite gain — any perturbation that chronically raises MAP above the equilibrium set point will be corrected to zero steady-state error by the pressure-natriuresis response. The only way to permanently raise MAP is to shift the pressure-natriuresis curve itself.*

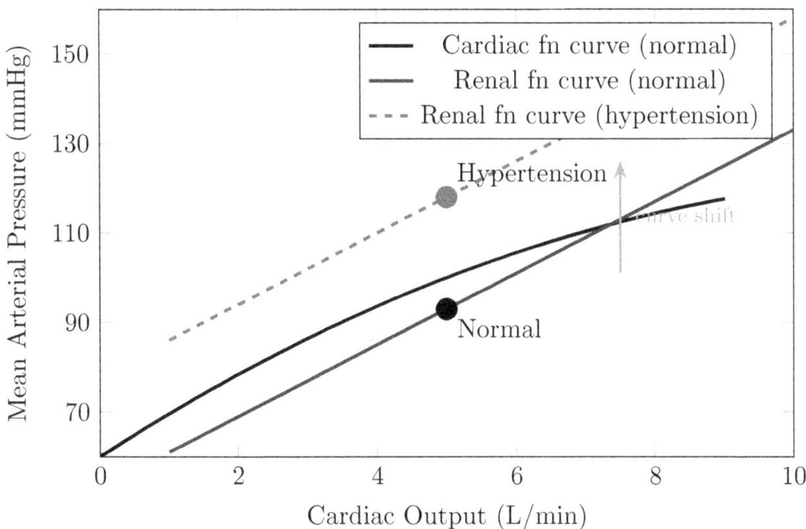

Figure 14.2: The Guyton diagram. The cardiac function curve (how CO determines MAP) intersects the renal function curve (how MAP determines volume, hence CO) at the equilibrium point. In hypertension, the renal function curve shifts upward by ≈ 25 mmHg, moving the equilibrium to a higher MAP at the same CO. Only interventions that shift the renal curve back down produce sustained blood pressure reduction.

14.5 The Clinical Interpretation: Vascular Remodelling and the Remodelled Resistance

Hypertension does not just raise the pressure setpoint. Over months to years it structurally remodels the vasculature in ways that reinforce the hypertension and resist treatment.

14.5.1 Vascular Remodelling: The Biology

What high pressure does to a blood vessel wall.

The arterial wall consists of three layers: the intima (endothelium and subendothelial matrix), the media (smooth muscle cells and elastin), and the adventitia (fibroblasts and collagen).

Chronic hypertension triggers three remodelling responses, each driven by specific molecular signals:

*1. **Medial hypertrophy.** Vascular smooth muscle cells (VSMCs) respond to chronic mechanical stretch by proliferating and synthesising more extracellular matrix. The signalling pathway: stretch $\rightarrow$ integrin activation $\rightarrow$ focal adhesion kinase (FAK) phosphorylation $\rightarrow$ ERK1/2 activation $\rightarrow$ VSMC proliferation. The result: the media thickens and the lumen narrows, increasing resistance even at the same vasomotor tone.*

*2. **Collagen deposition and arterial stiffening.** TGF-β (released by stretched VSMCs) drives fibroblast activation and collagen I/III deposition. This reduces arterial compliance C (Chapter 3), increasing pulse pressure and pulse wave velocity. Once collagen is deposited, it cannot be reversed by antihypertensive drugs — which is why long-standing hypertension leaves persistent arterial stiffness even after blood pressure is normalised.*

*3. **Endothelial dysfunction.** Chronic high shear stress and angiotensin II reduce eNOS expression via epigenetic methylation of the*

NOS3 promoter. Less NO production $\to$ less endothelium-dependent vasodilation $\to$ increased basal vasoconstriction. This creates a vicious cycle: hypertension reduces eNOS, reduced eNOS raises TPR, raised TPR maintains hypertension.

14.5.2   The Mathematics of Remodelling

Define the lumen radius of a remodelled vessel as $r = r_0(1 - \epsilon)$, where $\epsilon \in [0,1]$ is the fractional narrowing from medial hypertrophy. From Poiseuille (Chapter 2):

[Useful Approximation]

$$\mathcal{R}_{remodelled} = \frac{8\eta L}{\pi r_0^4 (1 - \epsilon)^4} = \frac{\mathcal{R}_0}{(1 - \epsilon)^4} \tag{14.6}$$

For $\epsilon = 0.1$ (10% lumen narrowing from wall thickening): resistance increases by $(1 - 0.1)^{-4} = (0.9)^{-4} = 1.52$ — 52% above baseline from structural remodelling alone.

This structural component of hypertension is why antihypertensive drugs produce a smaller blood pressure reduction than predicted from their pharmacological mechanism alone: the drugs reduce vasomotor tone (functional resistance) but cannot reverse the structural narrowing (structural resistance). Full reversal of hypertension requires both pharmacological treatment and time for the structural remodelling to partially reverse.

14.6   The Worked Example: Salt Sensitivity Assessment with Biological Variability

A 48-year-old woman with newly diagnosed hypertension undergoes assessment. Three separate measurements over 2 weeks (biological variability is stated explicitly):

Measurement	Value	Normal range
MAP (mean of 3)	112 mmHg	70–100 mmHg
Plasma renin	0.4 ng/mL/h	0.5–2.0 ng/mL/h
Aldosterone	22 ng/dL	3–16 ng/dL
ARR	55	<20
24h urine Na^+	210 mEq/day	100–200 mEq/day

Biological note: *MAP varies by $\pm$8–12 mmHg within a single day and by $\pm$5–8 mmHg between days in the same individual. The mean of three readings reduces measurement uncertainty to $\pm$4–7 mmHg — the quoted value is not a point estimate but the centre of a $\pm$5 mmHg range.*

Step 1 — Classify the hypertension. *Low renin (0.4 ng/mL/h) with high aldosterone (22 ng/dL) and ARR 55 ($>$ 20): primary aldosteronism is the most likely diagnosis (though the ARR threshold of 20 is laboratory- and assay-dependent; confirmatory testing with saline suppression or fludrocortisone is required before surgical planning). The kidney is retaining sodium under the influence of autonomous aldosterone secretion — a rightward shift in the pressure-natriuresis curve driven by mineralocorticoid excess.*

Step 2 — Quantify the setpoint shift. *From equation (14.3) with the measured 24h urine Na^+:*

The patient is in sodium balance at MAP 112 mmHg with excretion matching intake of 210 mEq/day. Normal $n \approx 3$. For her to be in balance at 93 mmHg (normal setpoint) she would need to excrete only 150 mEq/day (normal dietary intake). The excess 60 mEq/day at 93 mmHg represents the aldosterone-driven sodium retention. Her effective n at the operating point:

$$n_{\mathit{eff}} = \frac{\ln(210/150)}{\ln(112/93)} = \frac{0.336}{0.186} \approx 1.8$$

This is below the normal range of $n = 3$–5, confirming impaired (low-

slope) pressure-natriuresis — consistent with mineralocorticoid excess reducing the kidney's sodium-pressure sensitivity.

Step 3 — Treatment prediction. *If aldosterone-secreting adenoma is confirmed (adrenal CT, adrenal vein sampling), surgical removal or spironolactone will suppress aldosterone $\rightarrow$ shift the pressure-natriuresis curve leftward $\rightarrow$ restore n toward normal $\rightarrow$ achieve sodium balance at lower MAP. Predicted MAP reduction from spironolactone (based on the ARR and aldosterone level): 15–25 mmHg (range reflecting biological variability in adrenal responsiveness to mineralocorticoid receptor blockade).*

Why the ARR changes the entire treatment: *Essential hypertension (normal ARR) responds to almost any antihypertensive class. Primary aldosteronism (ARR $>$ 20) responds specifically to mineralocorticoid receptor antagonists or surgery. The same MAP of 112 mmHg has two completely different molecular causes and two completely different optimal treatments. The biology determines the mathematics, and the mathematics determines the treatment.*

Model Assumptions, Chapter 14. *What the Guyton model assumes: (1) Long-term MAP is set entirely by the kidney (Guyton's infinite gain principle). (2) The pressure-natriuresis curve is a power law with a single exponent n. (3) TPR and cardiac function shift the operating point acutely but not chronically. Where it breaks down: The infinite gain principle is an idealisation: real pressure-natriuresis curves have finite slope and show hysteresis. Structural vascular remodelling (not captured in the pressure-natriuresis equation) can maintain hypertension even after the renal setpoint is normalised. Individual n values vary 1–5-fold; the population-average approximation may be substantially wrong for a given patient.*

14.7 The Software Module

Module 14.1 — Hypertension Phenotyping and Pressure-Natriuresis Simulator

Input: MAP (mean of ≥ 3 readings, with SD); plasma renin activity; aldosterone; 24h urine Na^+; serum K^+; body weight/height; treatment if any.

Processing:

1. *Compute ARR; classify as low-renin (primary aldosteronism), normal-renin (essential), or high-renin (renovascular).*

2. *Estimate pressure-natriuresis slope n from operating point (equation 14.1).*

3. *Simulate pressure-natriuresis curves for normal and patient-specific n; plot Guyton diagram.*

4. *Predict MAP reduction for each drug class (ACEi/ARB, diuretic, CCB, MRA) based on renin-aldosterone phenotype and n.*

5. *Include biological variability bands (± 1 SD) on all predictions.*

Output: Hypertension classification; n estimate; pressure-natriuresis curve and Guyton diagram; treatment response prediction with uncertainty range; ARR interpretation.

Full implementation at themathematicsoftheliving-body.com/modules.

14.8   Chapter Summary

1. *Hypertension is a rightward shift of the pressure-natriuresis curve. The kidney retains sodium at a higher pressure setpoint, driven by molecular changes in Piezo1, eNOS, ENaC, or NKCC2 in the juxtaglomerular apparatus.*

2. *Guyton's model: long-term MAP is set entirely by the pressure-*

natriuresis relationship ($P^* = P_0[I_{Na}/\dot{U}^0_{Na}]^{1/n}$). The renal function curve's intersection with the cardiac function curve determines the equilibrium (Figure 14.2).

3. *Salt sensitivity is a small pressure-natriuresis slope exponent n. Its molecular basis includes NOS3 variants (reduced eNOS), ENaC overactivity, and impaired tubuloglomerular feedback.*

4. *Vascular remodelling (medial hypertrophy, collagen deposition, endothelial dysfunction) creates a structural resistance component that persists even after pharmacological treatment. Lumen narrowing of 10% raises Poiseuille resistance by 52%.*

5. *In the worked example (ARR 55), primary aldosteronism was identified as the mechanism. The effective $n \approx 1.8$ confirmed impaired pressure-natriuresis, and spironolactone was predicted to reduce MAP by 15–25 mmHg (range reflecting biological variability).*

6. *The biological and mathematical approaches are inseparable: the molecular phenotype (which cells, which proteins, which variants) determines n, which determines the optimal drug class. Biology writes the equation; the equation specifies the treatment.*

Key Equations, Chapter 14

$$P^* = P_0 \cdot \left(\frac{I_{Na}}{\dot{U}^0_{Na}} \right)^{1/n} \quad \textit{(long-term MAP)}$$

$$\mathcal{R}_{remod.} = \mathcal{R}_0 \cdot (1 - \epsilon)^{-4} \quad \textit{(remodelled resistance)}$$

$$n_{eff} = \frac{\ln(\dot{U}/\dot{U}_0)}{\ln(P/P_0)} \quad \textit{(Na-pressure sensitivity)}$$

$$ARR = \frac{[Aldo]}{PRA} \quad (> 20 \Rightarrow \textit{primary aldosteronism})$$

Key References, Chapter 14

1. *Hall, J.E. et al. (2012). Obesity-induced hypertension. Physiol. Rev. 92, 1081–1125. [Renal pressure-natriuresis mechanism in detail.]*

2. *Funder, J.W. et al. (2016). The Management of Primary Aldosteronism. J. Clin. Endocrinol. Metab. 101, 1889–1916. [ARR cutoffs and primary aldosteronism diagnosis used in worked example.]*

3. *Iadecola, C. & Davisson, R.L. (2008). Hypertension and cerebrovascular dysfunction. Cell Metab. 7, 476–484. [Vascular remodelling and eNOS epigenetics in hypertension.]*

Chapter 15

Shock: Collapse of the Control System

15.1 The Biology First: What Cells Experience During Shock

Before the equations: what happens inside a cell when oxygen delivery fails.

Shock is ultimately a cellular energy crisis. When oxygen delivery falls below cellular demand, mitochondrial oxidative phosphorylation cannot maintain the proton gradient across the inner mitochondrial membrane. ATP production from the electron transport chain ceases. Cells switch to anaerobic glycolysis — producing only 2 ATP per glucose versus 36 from aerobic metabolism.

The cascade of cellular failure unfolds in sequence:

1. Na^+/K^+-ATPase failure. This pump consumes $\approx$ 30% of resting cellular ATP. When ATP falls, pump activity decreases within minutes. Na^+ accumulates intracellularly, K^+ leaks out, and the cell membrane potential depolarises. Cells swell as water follows Na^+ osmotically.

2. Intracellular Ca^{2+} dysregulation. SERCA (sarcoplasmic/endoplasmic reticulum Ca^{2+}-ATPase) and plasma membrane Ca^{2+}-ATPase are also ATP-dependent. Without ATP, intracellular Ca^{2+} rises from its normal resting level of $\approx$ 100 nM to $>$ 1 μM. This triggers: activation of phospholipases (membrane damage), proteases (cytoskeletal destruction), and endonucleases (DNA fragmentation).

3. Mitochondrial membrane permeabilisation. High Ca^{2+} opens the mitochondrial permeability transition pore (mPTP), releasing cytochrome c into the cytosol. This triggers apoptosis cascades in some cells and necrosis in others depending on the remaining ATP level.

4. Lactic acidosis. Anaerobic glycolysis produces lactate and H^+. Blood lactate rises ($>$ 2 mmol/L indicates significant anaerobic metabolism; $>$ 4 mmol/L indicates severe shock). The acidosis impairs myocardial contractility (E_{es} falls), reduces catecholamine receptor sensitivity, and further inhibits mitochondrial function — a metabolic vicious cycle.

Why this matters for the mathematics: The critical threshold is not a MAP value. It is the point at which oxygen delivery (DO_2) falls below the critical oxygen delivery threshold DO_2^{crit} — below which cellular ATP production becomes supply-limited. This threshold, not

MAP, is what the treatment must target.

15.2 The Observation: Three Failure Modes of the Control System

Chapter 1 established $MAP = CO \times TPR$. Shock occurs when this equation can no longer be solved at $MAP \geq 65$ mmHg. Three distinct failure modes produce shock by attacking different terms:

Hypovolaemic shock: *Reduced $V_b \rightarrow$ reduced venous return $\rightarrow$ reduced CO. TPR rises reflexively. MAP falls when the CO fall exceeds the TPR rise.*

Cardiogenic shock: *Reduced $E_{es} \rightarrow$ reduced SV $\rightarrow$ reduced CO. TPR rises maximally. MAP falls when the pump is too weak even at maximum TPR compensation.*

Distributive shock (septic): *Massive vasodilation $\rightarrow$ reduced TPR. CO may be elevated (hyperdynamic), but MAP falls because TPR falls faster than CO rises. The control system error is in the vasculature, not the pump.*

Each failure mode has a different haemodynamic fingerprint:

Type	*CO*	*TPR*	*PCWP*	*SvO_2*
Hypovolaemic $\downarrow$		$\uparrow$	$\downarrow$	$\downarrow$
Cardiogenic $\downarrow$		$\uparrow$	$\uparrow$	$\downarrow$
Distributive $\uparrow$		$\downarrow$	$\downarrow$	$\uparrow$
Obstructive $\downarrow$		$\uparrow$	*Variable* $\downarrow$	

15.3 The Mathematics: The Oxygen Delivery Framework

15.3.1 Critical Oxygen Delivery

From Chapter 6, oxygen delivery: $DO_2 = CO \times C_a \times 10$ mL/min. Oxygen consumption $\dot{V}O_2$ is normally independent of DO_2 — the body extracts more oxygen when delivery falls, maintaining consumption via increased OER. This is supply-independent *oxygen consumption.*

Below a critical threshold DO_2^{crit}, extraction cannot increase further ($OER \approx 70\%$ is the maximum). Below this point, $\dot{V}O_2$ becomes supply-dependent — it falls proportionally with DO_2. This is the onset of cellular anaerobic metabolism and shock:

[Useful Approximation]

$$\dot{V}O_2 = \begin{cases} \dot{V}O_2^{normal} & DO_2 \geq DO_2^{crit} \\[2mm] \dot{V}O_2^{normal} \cdot \dfrac{DO_2}{DO_2^{crit}} & DO_2 < DO_2^{crit} \end{cases} \tag{15.1}$$

Normal $DO_2^{crit} \approx 300$ mL/min (range 250–350 mL/min, $\pm 20\%$ biological variability). This threshold is derived from population data and is not reliably measurable in individual patients — lactate elevation is the practical clinical surrogate. Normal resting $DO_2 \approx 1{,}000$ mL/min — providing a safety margin of $\approx$ 3-fold.

15.3.2 The Phase-Plane Analysis of Shock

Model the haemodynamic state as a two-dimensional system with state variables MAP and CO. Define the shock boundary as the set of (MAP, CO) pairs for which $DO_2 < DO_2^{crit}$:

[Core Principle]

$$DO_2 = CO \times C_a \times 10 < DO_2^{crit} \tag{15.2}$$

For fixed $C_a \approx 20$ mL/dL: $CO < DO_2^{crit}/200 = 300/200 = 1.5$ L/min

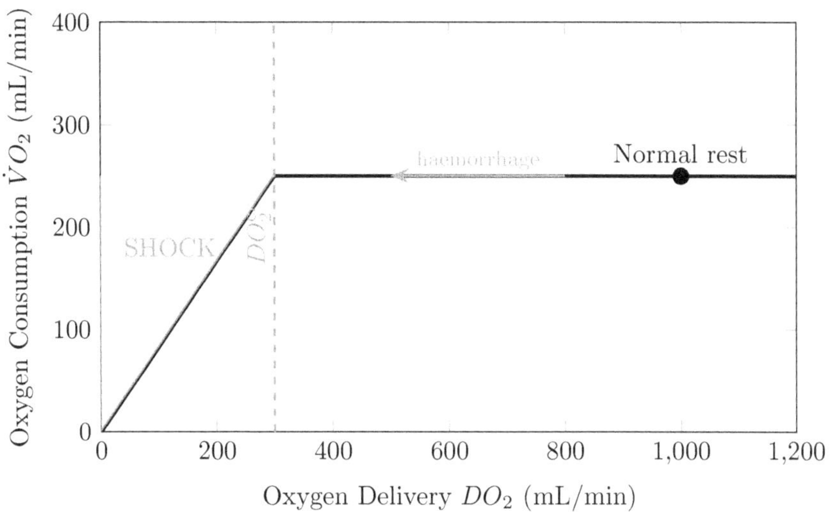

Figure 15.1: The oxygen delivery (DO_2) – consumption ($\dot{V}O_2$) relationship. Above DO_2^{crit} (≈ 300 mL/min, shaded boundary), $\dot{V}O_2$ is supply-independent: extraction increases to maintain consumption. Below DO_2^{crit}, $\dot{V}O_2$ falls proportionally — cellular anaerobic metabolism begins. Shock is defined by entry into the supply-dependent zone.

defines the critical CO threshold. The critical MAP depends on TPR:
$$MAP_{crit} = 1.5 \times TPR.$$

The decompensated shock transition is a bifurcation in the MAP-CO phase plane. The compensated state has high TPR, moderate CO, and MAP maintained near normal. The decompensated state has failed vasomotor tone, falling CO, and MAP below perfusion thresholds.

The transition is accelerated by the ischaemic myocardium: as MAP falls, coronary perfusion pressure falls, E_{es} falls (from acidosis and ischaemia), CO falls further, MAP falls further. This positive feedback loop is what makes shock irreversible once it passes the tipping point.

$$\frac{\mathrm{d}\,CO}{\mathrm{dt}} = \frac{1}{\tau_c}\left[f_{Frank\text{-}Starling}(EDV) \cdot E_{es}(MAP) - CO\right] \tag{15.3}$$

where $E_{es}(MAP)$ is the pressure-dependent contractility (Chapter 4), *which falls as MAP falls below the lower limit of coronary autoregula-tion (≈ 50 mmHg).*

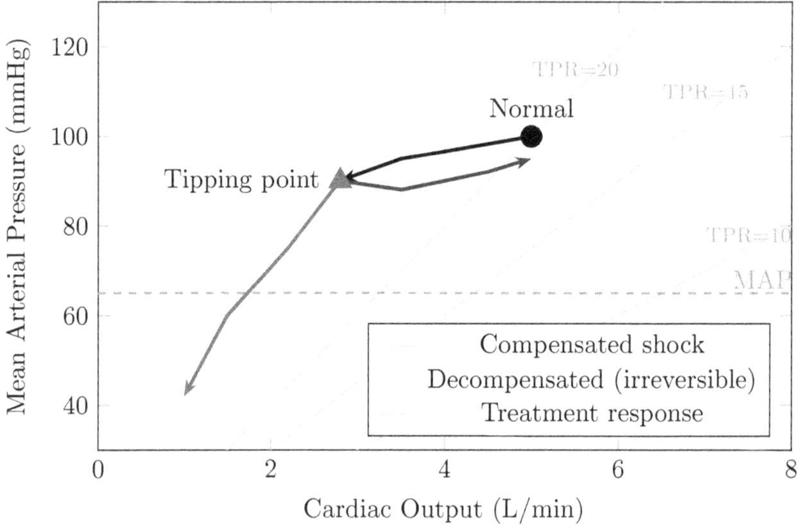

Figure 15.2: Phase plane of shock. Normal state (black dot, CO 5 L/min, MAP 100 mmHg). Compensated shock (black trajectory): haemorrhage reduces CO but TPR rises, maintaining MAP above 65 mmHg. At the tipping point (triangle), the feedback loop between falling MAP, falling coronary perfusion, falling E_{es}, and further falling MAP becomes irreversible (red trajectory). Early treatment (blue) reverses the trajectory before the tipping point is reached.

15.4   The Derivation: The Tipping Point (Irreversible Shock)

The tipping point between compensated and decompensated shock occurs at the MAP below which the coronary autoregulation lower limit is breached. From Chapter 11:

Below MAP $\approx$ 50–60 mmHg (with biological variability $\pm$10 mmHg depending on prior hypertensive remodelling):

$$E_{es}(MAP) \;=\; E_{es}^0 \cdot \frac{MAP^2}{MAP^2 + K_{1/2}^2} \tag{15.4}$$

where $K_{1/2} \approx 55$ mmHg is the MAP at half-maximal E_{es}. As MAP falls below 55 mmHg, E_{es} falls steeply, CO falls, MAP falls further — the decompensation loop.

The critical MAP (tipping point) satisfies $\mathrm{d}\,CO/\,\mathrm{d}\,MAP = 1/\,TPR$ (the rate of CO decline from ischaemia equals the rate of MAP decline it causes). Solving numerically with typical parameters gives: $MAP_{tipping} \approx$ 50–65 mmHg — consistent with the clinical observation that sustained $MAP < 65$ mmHg is associated with multi-organ failure.

> ### The molecular biology of the tipping point: why it is hard to reverse.
>
> *When MAP falls below ≈ 55 mmHg and the tipping point is crossed, three molecular events make reversal progressively harder:*
>
> ***1. iNOS induction.*** *Ischaemia and inflammatory cytokines (TNF-α, IL-1β) induce inducible nitric oxide synthase (iNOS, encoded by NOS2) in vascular smooth muscle and macrophages. Unlike eNOS (which produces small amounts of NO for signalling), iNOS produces massive sustained NO synthesis that causes pathological vasodilation and is resistant to vasopressors. This is why norepinephrine doses must be escalated in septic shock to overcome iNOS-driven vasodilation.*
>
> ***2. Mitochondrial uncoupling.*** *Reactive oxygen species (ROS) generated during ischaemia uncouple the mitochondrial electron transport chain from ATP synthesis by opening the mPTP. Even if oxygen delivery is restored, the damaged mitochondria cannot efficiently regenerate ATP. This is the cellular basis of* ischaemia-reperfusion injury *— cells that survive the shock period may die during reperfusion from ROS-driven damage.*
>
> ***3. Glycocalyx degradation.*** *The endothelial glycocalyx — a 0.5-*

1 µm layer of proteoglycans and glycoproteins lining the vascular lumen — regulates capillary permeability. Shock and inflammation degrade the glycocalyx (via matrix metalloproteinases and heparanase), causing capillary leak syndrome. Fluid administered to restore volume leaks into the interstitium rather than remaining intravascular, making volume resuscitation progressively less effective. This is why the initial litres of crystalloid work but subsequent litres do not.

15.5 The Clinical Interpretation: Matching Resuscitation to Failure Mode

Treatment of shock follows directly from the $MAP = CO \times TPR$ diagnostic framework. The goal is always to restore $DO_2 > DO_2^{crit}$, but the means differ by failure mode:

Shock type	*Primary fix*	*Drug/intervention*
Hypovolaemic	$\uparrow V_b$ *(restore CO)*	*Crystalloid, blood*
Cardiogenic	$\uparrow E_{es}$	*Dobutamine, MCS*
Distributive	$\uparrow TPR$	*Norepinephrine*
Obstructive	*Remove obstruction*	*Thrombolysis, drainage*

15.5.1 The Lactate Clearance Target

Blood lactate is the biochemical readout of $DO_2 < DO_2^{crit}$. Resuscitation adequacy is assessed by lactate clearance rate:

$$Clearance\ rate = \frac{[Lac]_0 - [Lac]_{2h}}{[Lac]_0} \times 100\% \qquad (15.5)$$

Target: $\geq 10\%$/hour clearance. If lactate fails to clear despite apparently adequate haemodynamics, consider: (1) inadequate microcirculatory flow *(macro-haemodynamics normal, micro-haemodynamics impaired); (2) mitochondrial dysfunction (cellular oxygen utilisation impaired despite adequate delivery); (3) ongoing occult haemorrhage.*

Why MAP $\geq$ 65 mmHg is not the same as adequate resuscitation: *MAP is a macrovascular parameter. Shock injures the microcirculation — the arterioles, capillaries, and venules where oxygen exchange actually occurs. After resuscitation, MAP may normalise while the microcirculation remains dysfunctional (heterogeneous flow, stopped-flow capillaries). Lactate and $ScvO_2$ are the biochemical readouts of cellular oxygen adequacy; MAP alone is insufficient. This is the mathematical distinction between perfusion pressure (a macrovascular variable) and oxygen delivery (a cellular variable). Both must be targeted.*

15.6 The Worked Example: Haemorrhagic Shock with Biological Variability

A 34-year-old man sustains a stab wound to the abdomen. On arrival: HR 128 bpm (range 118–138 in two readings), MAP 64 mmHg (58–71), GCS 13/15, cool extremities, urine output 18 mL/h. Haemoglobin 9.4 g/dL (measured twice: 9.1 and 9.7 g/dL). Lactate 4.8 mmol/L (range 4.2–5.4 on point-of-care with $\pm 10\%$ analytical error).

Step 1 — Classify and estimate DO_2.

$$C_a = 1.34 \times 9.4 \times 0.98 \approx 12.3 \ mL/dL \quad (range \ 11.8\text{–}12.9)$$

$$CO_{est} \approx \frac{MAP}{TPR_{assumed}} \approx \frac{64}{22} \approx 2.9 \ L/min$$

$$DO_2 = 2.9 \times 12.3 \times 10 \approx 357 \ mL/min \quad (range \ 300\text{–}420)$$

DO_2 is near DO_2^{crit} (300–350 mL/min). With the lower bound of the estimate, this patient may already be in the supply-dependent zone — confirmed by lactate $= 4.8$ mmol/L (> 4 mmol/L $=$ severe anaerobic metabolism).

Step 2 — Treatment targets. *Target $DO_2 > 450$ mL/min (safety margin above DO_2^{crit}). To achieve this with $C_a \approx 12.3$ mL/dL:*

$$CO_{needed} = \frac{450}{12.3 \times 10} \approx 3.7 \ L/min$$

Increase CO from 2.9 to 3.7 L/min — a 28% increase. Primary means: restore V_b with blood transfusion (target $Hb \geq 10$ g/dL) and control haemorrhage surgically.

Step 3 — Track lactate clearance. *Measure lactate at 0 and 2 hours. Target $\geq 20\%$ clearance in 2 hours. If lactate rises or fails to clear despite restoration of MAP > 65 mmHg, escalate search for ongoing haemorrhage.*

Biological variability note: *DO_2^{crit} varies 250–350 mL/min between individuals (lower in trained athletes with efficient mitochondria; higher in patients with impaired cellular oxygen utilisation from sepsis or mitochondrial disease). The treatment target should be lactate-guided rather than fixed-DO_2-guided for this reason.*

Model Assumptions, Chapter 15. *What the shock model assumes: (1) DO_2^{crit} is a sharp threshold (in reality, it is a gradual transition zone with $\pm 20\%$ inter-individual variability). (2) Cardiac output is the limiting factor (not microcirculatory distribution of flow). (3) The tipping point can be predicted from MAP alone. Where it breaks down: Microcirculatory failure can occur even when macrovascular MAP and CO are restored — the so-called "microcirculatory-macrocirculatory dissociation" of sepsis. Lactate may be elevated from causes other than anaerobic metabolism (thiamine deficiency, liver failure, epinephrine infusion). The MAP tipping point ($\approx$ 50–65 mmHg) varies with individual autoregulatory capacity.*

15.7   The Software Module

Module 15.1 — Shock Classifier and Resuscitation Target Calculator

Input: *HR, MAP (mean $\pm$ SD of readings); haemoglobin (g/dL); SaO_2 (%); SvO_2 or $ScvO_2$ (%); PCWP (if available); lactate (mmol/L); urine output (mL/h); temperature ($^\circ C$).*

Processing:

1. *Compute C_a, DO_2, OER from Fick equations (Module 6.1).*

2. *Classify shock type from CO/PCWP/TPR pattern (Chapter 1 framework).*

3. *Compute distance from DO_2^{crit} (equation 15.1); flag supply-dependent $\dot{V}O_2$ if OER > 60% or lactate > 2 mmol/L.*

4. *Compute treatment target CO and V_b to reach DO_2 > 450 mL/min.*

5. *Track lactate clearance at 1 and 2 h; flag inadequate clearance ($< 10\%/h$).*

6. *Include biological variability bands on all computed quantities.*

Output: *Shock classification; DO_2 with uncertainty range; distance to DO_2^{crit}; treatment targets (CO, Hb, MAP); lactate clearance rate; tipping point proximity estimate.*

Full implementation at themathematicsoftheliving-body.com/modules.

15.8 Chapter Summary

1. *Shock is cellular energy failure, not low blood pressure. The defining event is $DO_2 < DO_2^{crit} \approx 300$ mL/min (range 250–350, $\pm 20\%$ biological variability), below which $\dot{V}O_2$ becomes supply-dependent and anaerobic metabolism begins.*

2. *The cellular cascade: ATP depletion $\rightarrow$ Na^+/K^+-ATPase failure $\rightarrow$ cell swelling $\rightarrow$ Ca^{2+} dysregulation $\rightarrow$ mitochondrial permeabilisation $\rightarrow$ necrosis or apoptosis. Lactic acidosis impairs E_{es} and creates a metabolic vicious cycle.*

3. *The tipping point between compensated and irreversible shock occurs at MAP $\approx$ 50–65 mmHg, where coronary autoregulation fails, E_{es} falls, and a positive feedback loop drives progressive circulatory collapse.*

4. *iNOS induction (massive sustained NO), mitochondrial uncoupling (ischaemia-reperfusion injury), and glycocalyx degradation (capillary leak) are the three molecular mechanisms that make shock increasingly treatment-resistant once the tipping point is crossed.*

5. *Treatment targets the specific failure mode: volume ($\uparrow V_b$) for hypovolaemic, inotropes ($\uparrow E_{es}$) for cardiogenic, vasopressors ($\uparrow TPR$) for distributive, obstruction relief for obstructive shock.*

6. *Lactate clearance ($\geq$ 10%/h) is the biochemical target of resuscitation — superior to MAP alone because it reflects cellular oxygen adequacy rather than macrovascular perfusion pressure.*

Key Equations, Chapter 15

$$DO_2 = CO \times C_a \times 10$$

$$\dot{V}O_2 = \dot{V}O_2^{norm} \cdot \min\left(1, \frac{DO_2}{DO_2^{crit}}\right)$$

$$E_{es}(MAP) = E_{es}^0 \cdot \frac{MAP^2}{MAP^2 + K_{1/2}^2}$$

$$\text{Lactate clearance} = \frac{[Lac]_0 - [Lac]_{2h}}{[Lac]_0} \times 100\%$$

Key References, Chapter 15

1. Vincent, J.L. et al. (1995). *Use of the SOFA score to assess the incidence of organ dysfunction/failure in intensive care units.* Crit. Care Med. 23, 1793–1800. [SOFA score for clinical shock assessment.]

2. Kruse, J.A. et al. (1987). *The relationship between oxygen delivery*

and consumption during fluid resuscitation of hypovolemic and septic shock. Chest 92, 383–391. [DO_2-$\dot{V}O_2$ relationship and critical threshold.]

3. Ince, C. (1998). The microcirculation is the motor of sepsis. Crit. Care 9 (Suppl 4), S13–S19. [Microcirculatory failure beyond macrovascular resuscitation.]

Chapter 16

Atherosclerosis: Wall Shear Stress and Plaque

16.1 The Biology First: The Endothelial Cell as a Mechanosensor

Before the equations: how the endothelium reads the flow field.

The arterial endothelium is not a passive lining. It is an active mechanosensory organ that continuously reads the mechanical forces applied to it by flowing blood and responds with gene expression changes that either protect or harm the vessel wall.

The primary mechanical signal is wall shear stress *(WSS) — the tangential force per unit area exerted by flowing blood on the endothelial surface. Endothelial cells possess at least four classes of mechanosensors that detect WSS:*

*1. **PECAM-1/VE-cadherin/VEGFR2 complex.** This transmembrane protein complex at endothelial junctions acts as a force transducer. Shear stress causes PECAM-1 to undergo conformational change, triggering PI3K activation, Akt phosphorylation, and downstream transcription factor activation. High, steady, laminar WSS (10–30 dyn/cm^2) activates this pathway to produce atheroprotective outputs (eNOS upregulation, anti-inflammatory gene expression). Low or oscillatory WSS (< 4 dyn/cm^2) activates NF-κB and AP-1 instead, upregulating adhesion molecules (VCAM-1, ICAM-1), chemokines (MCP-1), and pro-coagulant factors.*

*2. **The glycocalyx.** The 0.5–1 μm glycocalyx (heparan sulphate proteoglycans, hyaluronan, sialoproteins) is the primary WSS sensor. Shear stress deforms the glycocalyx, transmitting force via its core proteins to the cytoskeleton and triggering Ca^{2+} signalling and eNOS activation. Glycocalyx disruption (by oxLDL, hyperglycaemia, or inflammation) abolishes shear-mediated eNOS activation even when WSS is normal — decoupling the mechanical signal from the biological response.*

*3. **Ion channels.** Piezo1 (Chapter 14) and K$_{Ca}$ (large conductance Ca^{2+}-activated K$^+$ channels) are mechanosensitive ion channels in the endothelial membrane. Shear stress opens these channels, causing Ca^{2+} influx and membrane hyperpolarisation, both of which activate eNOS.*

*4. **Primary cilia.** Endothelial primary cilia (single, non-motile organelles) deflect in flow and signal through PC1/PC2 (polycystin 1/2) channels. They are particularly important in detecting pulsatile* *versus steady flow.*

The critical distinction: *High laminar WSS $\to$ eNOS $\to$ NO $\to$ atheroprotection. Low/oscillatory WSS $\to$ NF-κB $\to$ VCAM-1, MCP-1 $\to$ monocyte adhesion $\to$ foam cell formation $\to$ atherosclerosis.*

The spatial distribution of WSS in the arterial tree is determined entirely by the fluid mechanics. This is why atherosclerosis is a disease of geometry.

16.2 The Observation: Where Plaques Form

Atherosclerotic plaques do not form uniformly. They have a precise anatomical distribution that mirrors the low-WSS regions of the arterial tree:

- **Carotid bifurcation:** *The outer wall of the carotid bulb, where flow separates and recirculates. The inner wall (high WSS) is protected.*

- **Coronary artery bends:** *The outer curvature of coronary artery bends experiences low WSS from flow separation.*

- **Aortic bifurcation:** *The lateral walls of the iliac bifurcation receive low WSS from the splitting jet.*

- **Distal to stenoses:** *Once a plaque forms and creates a stenosis, the post-stenotic recirculation zone creates additional low-WSS regions, accelerating plaque progression downstream.*

The inner curvatures and straight segments of arteries — where WSS is high and laminar — are protected. The geometry writes the pathology.

16.3 The Mathematics: Wall Shear Stress from Poiseuille

16.3.1 WSS from the Velocity Profile

From Chapter 2, the parabolic velocity profile in a straight tube is: $v(r) = \frac{\Delta P}{4\eta L}(R^2 - r^2)$. *Wall shear stress is the viscous force per unit area at the wall* $(r = R)$:

$$\tau_w = \eta \left.\frac{\mathrm{d}v}{\mathrm{d}r}\right|_{r=R} = \eta \cdot \frac{-2\Delta P R}{4\eta L} \cdot (-1) = \frac{\Delta P R}{2L} \tag{16.1}$$

Substituting Poiseuille's equation $(\Delta P = 8\eta L Q/\pi R^4)$:

$$\tau_w = \frac{4\eta Q}{\pi R^3} \tag{16.2}$$

Wall shear stress depends on R^{-3} *(not* R^{-4} *as for resistance). A 20% reduction in radius increases WSS by* $(0.8)^{-3} = 1.95$*-fold.*

At physiological flow rates in the coronary artery $(Q \approx 1 \ mL/s, \ R = 1.5 \ mm, \ \eta = 3 \times 10^{-3} \ Pa{\cdot}s)$:

$$\tau_w = \frac{4 \times 3 \times 10^{-3} \times 10^{-6}}{\pi \times (1.5 \times 10^{-3})^3} = \frac{1.2 \times 10^{-8}}{1.06 \times 10^{-8}} \approx 1.13 \ Pa = 11.3 \ dyn/cm^2$$

This is in the atheroprotective range $(> 10 \ dyn/cm^2)$. *In the carotid bulb, where flow recirculates:* $\tau_w \approx 1\text{–}4 \ dyn/cm^2$ *— the pro-atherogenic range.*

16.3.2 Oscillatory Shear Index

In regions of flow separation, WSS oscillates in direction as well as magnitude. The oscillatory shear index (OSI) quantifies this:

$$OSI = \frac{1}{2} \left(1 - \frac{\left| \int_0^T \tau_w \, \mathrm{d}t \right|}{\int_0^T |\tau_w| \, \mathrm{d}t} \right) \qquad (16.3)$$

$OSI = 0$: purely unidirectional shear (atheroprotective). $OSI = 0.5$: perfectly oscillatory (maximum pro-atherogenic signal). High OSI regions strongly correlate with plaque localisation.

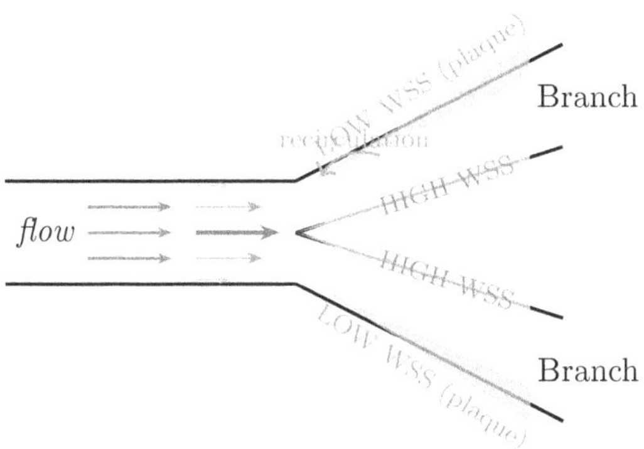

Figure 16.1: Wall shear stress distribution at an arterial bifurcation. The outer walls of both branches (red shading) experience low WSS due to flow separation and recirculation — the pro-atherogenic environment where VCAM-1 and MCP-1 are upregulated. The inner walls (green shading) experience high WSS from the accelerated jet — atheroprotected via eNOS upregulation. Plaques form on the outer walls, not the inner.

16.4 The Derivation: Plaque Growth as a Reaction-Diffusion Problem

16.4.1 The Intimal Thickening ODE

Atherosclerosis begins with lipid accumulation in the intima. The rate of LDL infiltration into the subendothelial space depends on the permeability of the endothelium, which is regulated by WSS:

$$J_{LDL} = L_p(\tau_w) \cdot \sigma(\tau_w) \cdot [LDL]_{plasma} \qquad (16.4)$$

where $L_p(\tau_w)$ is the hydraulic conductivity (increased at low WSS, as tight junctions open) and $\sigma(\tau_w)$ is the reflection coefficient (decreased at low WSS). At low WSS ($\tau_w < 4$ dyn/cm^2): $L_p \approx 3 \times 10^{-8}$ cm/s/mmHg; at high WSS: $L_p \approx 5 \times 10^{-9}$ cm/s/mmHg — a 6-fold difference in permeability from WSS alone.

Intimal thickness $h(t)$ grows as the sum of LDL infiltration, foam cell formation, and smooth muscle migration:

$$\frac{dh}{dt} = k_{foam} \cdot J_{LDL}(\tau_w) \cdot [oxLDL] - k_{regress} \cdot h \qquad (16.5)$$

where k_{foam} is the foam cell formation rate constant and $k_{regress}$ captures regression processes. At steady state:

$$h^* = \frac{k_{foam} \cdot J_{LDL}(\tau_w) \cdot [oxLDL]}{k_{regress}} \qquad (16.6)$$

The equilibrium plaque thickness is proportional to LDL flux (which depends on WSS) and oxLDL concentration (which depends on systemic LDL and oxidative stress). This is why lowering LDL reduces plaque growth even without changing WSS — it reduces the driving force $J_{LDL} \cdot [oxLDL]$ directly.

16.4.2 Plaque Vulnerability: The Cap Stress

Not all plaques are equally dangerous. A thin fibrous cap overlying a large lipid core is vulnerable to rupture. Cap stress is calculated from the modified Laplace equation:

$$\sigma_{cap} = \frac{P \cdot r_{lumen}}{2h_{cap}} \qquad (16.7)$$

where P is luminal pressure, r_{lumen} is the lumen radius at the plaque, and h_{cap} is the fibrous cap thickness. Rupture occurs when σ_{cap} exceeds the *tensile strength of the cap collagen (≈ 300 kPa).*

For a typical vulnerable plaque: $P \approx 120$ mmHg $= 16{,}000$ Pa, $r_{lumen} \approx 1.5$ mm, $h_{cap} \approx 65$ μm:

$$\sigma_{cap} = \frac{16{,}000 \times 1.5 \times 10^{-3}}{2 \times 65 \times 10^{-6}} = \frac{24}{1.3 \times 10^{-4}} \approx 185 \ kPa$$

This is below the rupture threshold (300 kPa), but during a hypertensive surge ($P = 200$ mmHg):

$$\sigma_{cap} \approx 185 \times \frac{200}{120} \approx 308 \ kPa > 300 \ kPa \ (rupture)$$

This is the mathematical basis for the clinical observation that acute plaque rupture is triggered by sudden pressure surges — morning hypertension, emotional stress, cocaine use, heavy exertion.

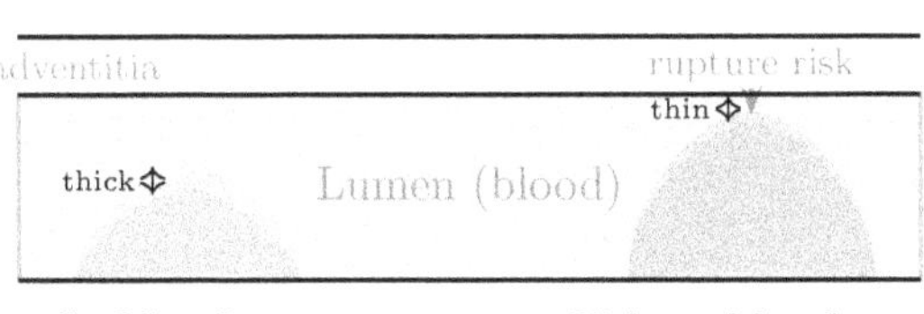

Figure 16.2: Stable versus vulnerable plaque. The stable plaque (left) has a thick fibrous cap (grey) overlying a modest lipid core (orange). Cap stress is low. The vulnerable plaque (right) has a thin fibrous cap over a large lipid core (red). A hypertensive surge raises cap stress above the 300 kPa rupture threshold, triggering acute MI or stroke. $\sigma_{\text{cap}} = PR/2h_{\text{cap}}$.

16.5 The Molecular Biology of Foam Cell Formation

> **The molecular cascade from low WSS to foam cell.**
>
> *Low WSS activates NF-κB in endothelial cells, which upregulates VCAM-1 and ICAM-1 on the endothelial surface. Circulating monocytes express $\alpha_4\beta_1$ integrin (VLA-4), which binds VCAM-1 — the molecular handshake that initiates monocyte adhesion and transmigration into the subendothelial space.*
>
> *Once in the intima, monocytes differentiate into macrophages under the influence of M-CSF (macrophage colony-stimulating factor). These macrophages express scavenger receptors (SR-A1, SR-B2, CD36) that bind oxidised LDL (oxLDL) via charge-charge interactions — unlike the regulated LDL receptor that downregulates with cholesterol loading, scavenger receptors are not feedback-inhibited. Macrophages continue ingesting oxLDL until they become cholesterol-laden* foam cells.
>
> **The mathematical consequence:** *The rate of foam cell formation in equation (16.5) depends on [oxLDL] — not [LDL]. Statins lower plasma LDL, reducing the substrate for oxidation and thereby reducing [oxLDL]. But statins also have direct anti-inflammatory effects (reducing VCAM-1, reducing macrophage scavenger receptor expression) that are independent of lipid lowering. This* pleiotropic *statin effect explains why statins reduce cardiovascular events even in patients with near-normal LDL.*

16.6 The Clinical Interpretation: Risk Stratification from Fluid Mechanics

16.6.1 Coronary Fractional Flow Reserve (FFR)

FFR (Chapter 2) quantifies whether a specific stenosis reduces coronary flow enough to cause ischaemia. But FFR does not identify vulnerable plaques. A haemodynamically non-significant stenosis (FFR > 0.8) may still have a thin fibrous cap and high rupture risk.

The combination of WSS mapping (from coronary CT angiography with computational fluid dynamics) and cap thickness assessment (from optical coherence tomography, OCT) provides a mechanistic risk stratification:

WSS	Cap thick.	Clinical risk
High (>15 dyn/cm^2)	Thick (>100 μm)	Stable, low rupture
Low (<4 dyn/cm^2)	Thick (>100 μm)	Growing, not yet vulnerable
High (>15 dyn/cm^2)	Thin (<65 μm)	High rupture risk
Low (<4 dyn/cm^2)	Thin (<65 μm)	Highest risk: growing and vulnerable

Why most MIs occur from non-obstructive plaques: *Autopsy studies show that $\approx$ 65% of fatal MIs arise from plaques that caused < 50% stenosis before rupture. The haemodynamically significant stenosis (FFR < 0.8) is the plaque that causes angina. The vulnerable plaque (thin cap, large lipid core, low WSS microenvironment) is the plaque that causes MI. These are not the same lesion. The cap stress equation $\sigma_{cap} = PR/2h_{cap}$ identifies the dangerous one. The flow equation identifies the symptomatic one. Both are needed.*

16.7 The Worked Example: Computing WSS and Cap Stress in a Patient Lesion

A 62-year-old man with stable angina undergoes coronary CT angiography. A lesion is identified in the LAD: lumen radius 1.2 mm (range 1.0–1.4 on repeat measurement), resting coronary flow estimated $Q = 0.7$ mL/s (range 0.5–1.0 from coronary flow reserve data). OCT cap thickness at the minimum: 82 μm (range 68–96).

Step 1 — Compute WSS.

$$\tau_w = \frac{4\eta Q}{\pi R^3} = \frac{4 \times 3 \times 10^{-3} \times 7 \times 10^{-7}}{\pi \times (1.2 \times 10^{-3})^3}$$

$$= \frac{8.4 \times 10^{-9}}{5.43 \times 10^{-9}} \approx 1.55 \ Pa = 15.5 \ dyn/cm^2$$

With the lower-bound flow $(Q = 5 \times 10^{-7} \ m^3/s)$*:* $\tau_w \approx 11.0 \ dyn/cm^2$. *Range:* **11–15.5 dyn/cm²** *(atheroprotective range,* $> 10 \ dyn/cm^2$*).*

Step 2 — Compute cap stress. *At resting MAP 105 mmHg =* *14,000 Pa:*

$$\sigma_{cap} = \frac{14{,}000 \times 1.2 \times 10^{-3}}{2 \times 82 \times 10^{-6}} = \frac{16.8}{1.64 \times 10^{-4}} \approx 102 \ kPa$$

Range (min cap, max pressure 140 mmHg = 18,700 Pa, min cap 68 μm): $\sigma_{max} = 18{,}700 \times 1.2 \times 10^{-3}/(2 \times 68 \times 10^{-6}) \approx 165 \ kPa.$

Step 3 — Risk stratification. *WSS is atheroprotective (high). Cap stress at rest is 102 kPa (safe,* $\ll 300 \ kPa$ *threshold). Even at hypertensive surge (200 mmHg), predicted cap stress* $\approx 195 \ kPa$ *— still below rupture threshold. This lesion is* **stable***.*

Contrast*: if cap thickness were 45 μm (thin cap):* $\sigma_{cap} = 14{,}000 \times 1.2 \times 10^{-3}/(2 \times 45 \times 10^{-6}) = 187 \ kPa$ *at rest; at 200 mmHg surge: 280 kPa. Approaching the rupture threshold — this would be classified as vulnerable.*

> ***Model Assumptions, Chapter 16.*** *What the WSS/plaque model assumes: (1) Blood behaves as a Newtonian fluid at the arterial wall (WSS calculated from Poiseuille, which assumes Newtonian flow). (2) The plaque cap is isotropic and homogeneous. (3) Cap stress is uniform around the circumference. Where it breaks down: Red blood cells migrate away from the vessel wall (Fåhræus-Lindqvist effect), creating a plasma-rich layer that reduces apparent near-wall viscosity. Real cap stress has stress concentrations at the cap shoulder (junction with normal wall) that far exceed the mean Laplace calculation. The 300 kPa rupture threshold is a population average with ±50% biological variability in cap material properties.*

16.8 The Software Module

Module 16.1 — WSS and Plaque Vulnerability Calculator

Input: *Lumen radius (mm, with measurement SD); coronary flow (mL/s, with physiological range); cap thickness from OCT (μm, with SD); resting and peak MAP; blood viscosity (default 3×10^{-3} Pa·s).*

Processing:

1. *Compute WSS (equation 16.2) with uncertainty range from input SDs.*

2. *Compute OSI estimate for bifurcation geometry (if applicable).*

3. *Compute cap stress (equation 16.7) at resting and peak MAP.*

4. *Classify WSS range (atheroprotective > 10, atherogenic < 4 dyn/cm^2).*

5. *Classify vulnerability: stable / growing / vulnerable / highest-risk using 2×2 table.*

6. *Compute critical MAP for rupture: $P_{crit} = 300 \times 2h_{cap}/r_{lumen}$ with uncertainty band.*

Output: *WSS with range; cap stress at rest and peak; vulnerability classification; critical MAP for rupture with uncertainty.*

Full implementation at themathematicsoftheliving-body.com/modules.

16.9 Chapter Summary

1. *Atherosclerosis is a disease of geometry. Low WSS (< 4 dyn/cm^2) at bends and bifurcations activates NF-κB in endothelial cells via the PECAM-1/glycocalyx/ Piezo1 mechanosensing apparatus, upregulating VCAM-1 and initiating monocyte adhesion.*

2. *WSS from Poiseuille: $\tau_w = 4\eta Q/\pi R^3$. WSS scales as R^{-3} (not R^{-4} for resistance). A 20% radius reduction approximately doubles WSS.*

3. *Plaque growth depends on LDL flux (J_{LDL}, increased 6-fold at low WSS) and oxLDL concentration. Statins reduce both the substrate and the macrophage scavenger receptor response — explaining their pleiotropic cardiovascular benefit beyond LDL lowering.*

4. *Cap stress $\sigma_{cap} = PR/2h_{cap}$ predicts rupture risk. Thin caps ($<$ 65 μm) rupture during hypertensive surges ($>$ 160 mmHg) when cap stress exceeds 300 kPa.*

5. *65% of fatal MIs arise from non-obstructive plaques (FFR $>$ 0.8). The haemodynamically significant plaque and the vulnerable plaque are not the same lesion. Both WSS (geometry) and cap stress (vulnerability) are needed for complete risk stratification.*

6. *In the worked example: WSS 11–15 dyn/cm^2 (atheroprotective), cap stress 102 kPa at rest (stable), rising to $<$ 200 kPa at hypertensive surge — classified as stable, no immediate rupture risk.*

Key Equations, Chapter 16

$$\tau_w = \frac{4\eta Q}{\pi R^3} \quad \text{(wall shear stress)}$$

$$OSI = \frac{1}{2}\left(1 - \frac{\left|\int_0^T \tau_w \, dt\right|}{\int_0^T |\tau_w| \, dt}\right)$$

$$\sigma_{cap} = \frac{Pr}{2h_{cap}} \quad \text{(plaque rupture stress)}$$

$$h^* = \frac{k_{foam}J_{LDL}[oxLDL]}{k_{regress}} \quad \text{(equilibrium plaque thickness)}$$

Key References, Chapter 16

1. *Caro, C.G. et al. (1971). Arterial wall shear and distribution of early atheroma in man. Nature 223, 1159–1161. [Original demonstration*

of WSS-atherosclerosis localisation in humans.]

2. Chatzizisis, Y.S. et al. (2007). Role of endothelial shear stress in the natural history of coronary atherosclerosis. J. Am. Coll. Cardiol. 49, 2379–2393. [WSS and plaque progression review.]

3. Falk, E. et al. (1995). Coronary plaque disruption. Circulation 92, 657–671. [Fibrous cap thickness and plaque vulnerability — basis of the 65 μm threshold.]

Chapter 17

Drug Mathematics: How Cardiovascular Drugs Move the Parameters

A drug does not treat a disease. It binds to a protein and changes its activity by a calculable amount. Everything downstream — the haemodynamic response, the side effect, the clinical outcome — follows from that single molecular event and the network it perturbs.

— Zachariah Sinkala

17.1 The Biology First: Drugs as Protein Modifiers

Before the equations: what a cardiovascular drug actually does at the molecular level.

Every cardiovascular drug works by binding to a specific protein and either inhibiting or activating it. The mathematical consequence is a change in one or more parameters of the cardiovascular ODE system developed in Chapters 1–16. Understanding which protein, how it is bound, and what the binding does to its activity is the biological foundation of drug mathematics.

__Competitive inhibition__ (most receptor antagonists and enzyme inhibitors): the drug competes with the endogenous ligand for the same binding site. The inhibition follows the Michaelis-Menten competition model:

$$\frac{v_{drug}}{v_{max}} = \frac{[S]}{[S] + K_m(1 + [I]/K_i)}$$

where $[I]$ is the drug concentration and K_i is the inhibition constant. At therapeutic plasma concentrations, $[I]/K_i$ determines the degree of inhibition.

__Inverse agonism__ (many β-blockers): the drug not only blocks the receptor but actively stabilises its inactive conformation, producing effects below basal activity. This is why β-blockers reduce resting heart rate below its intrinsic SAN rate.

__Allosteric modulation__ (ivabradine, digoxin): the drug binds outside the active site and changes the protein's conformation, altering its activity without directly competing with the endogenous ligand.

__The pharmacokinetic-pharmacodynamic (PK/PD) link__: the drug concentration at the receptor $[I]$ is not the plasma concentration. It is the free (unbound) fraction of the plasma concentration, adjusted for tissue distribution (V_d) and protein binding (typically 90–99% for cardiovascular drugs). The PK model predicts $[I](t)$; the PD model translates $[I]$ into fractional receptor occupancy; the CV ODE translates occupancy into haemodynamic effect.

17.2 The Observation: One Drug, One Parameter

The central organising principle of this chapter is that each cardiovascular drug class acts primarily on one parameter of the $MAP = CO \times TPR$ framework. The haemodynamic response is predictable from that parameter change, subject to the compensatory responses of the baroreflex (Chapter 9) and RAAS (Chapter 10).

Drug class	Primary target	Parameter changed
β-blocker	β_1 receptor	$\downarrow$ HR, $\downarrow$ E_{es} acutely
CCB (dihydropyridine)	L-type Ca^{2+} channel	$\downarrow$ TPR (vasodilation)
CCB (non-DHP)	L-type Ca^{2+} channel	$\downarrow$ HR, $\downarrow E_{es}$
ACE inhibitor	ACE enzyme	$\downarrow P_{AO}$, $\downarrow$[Aldo]
ARB	AT_1 receptor	$\downarrow P_{AO}$, $\downarrow$[Aldo]
Thiazide diuretic	NCC cotransporter	$\downarrow V_b$, $\downarrow$ EDV
Loop diuretic	NKCC2 cotransporter	$\downarrow V_b$, acute
Digoxin	Na/K-ATPase	$\uparrow E_{es}$, $\downarrow$ HR
Nitrates	sGC $\rightarrow$ cGMP	$\downarrow$EDV (venodilatation)
Ivabradine	HCN4 (I_f)	$\downarrow$ HR only

17.3 The Mathematics: The Dose-Response and the Hill Equation

17.3.1 Receptor Occupancy and Drug Effect

The fraction of receptors occupied by drug D at concentration $[D]$ follows the Hill equation:

$$\theta([D]) \;=\; \frac{[D]^n}{EC_{50}^n + [D]^n} \tag{17.1}$$

where EC_{50} is the drug concentration for 50% maximal effect and n is the Hill coefficient (cooperativity; $n = 1$ for most receptors, $n > 1$ indicates cooperativity).

The pharmacodynamic effect E is related to occupancy by:

$$E([D]) \;=\; E_0 + (E_{\max} - E_0) \cdot \theta([D]) \tag{17.2}$$

where E_0 is the baseline effect (no drug) and $E_{\max}$ is the maximum effect (all receptors occupied). For a β-blocker: $E_0 = HR_{baseline}$, $E_{\max} = HR_{\min}$ (minimum achievable HR at full β_1 blockade).

17.3.2 PK/PD Integration: The Effect Compartment

For most cardiovascular drugs, the plasma concentration $C_p(t)$ is not in equilibrium with the effect-site concentration $C_e(t)$ — there is a delay as the drug distributes to its tissue target. The effect compartment model:

$$\frac{\mathrm{d}C_e}{\mathrm{d}t} \;=\; k_{e0}(C_p - C_e) \tag{17.3}$$

where k_{e0} is the equilibration rate constant $(t_{1/2,eq} = \ln 2/k_{e0})$. For IV esmolol ($\beta$-blocker with rapid offset): $t_{1/2,eq} \approx 2$ min. For oral metoprolol: $t_{1/2,eq} \approx 20$ min. The haemodynamic effect tracks $C_e(t)$, not $C_p(t)$.

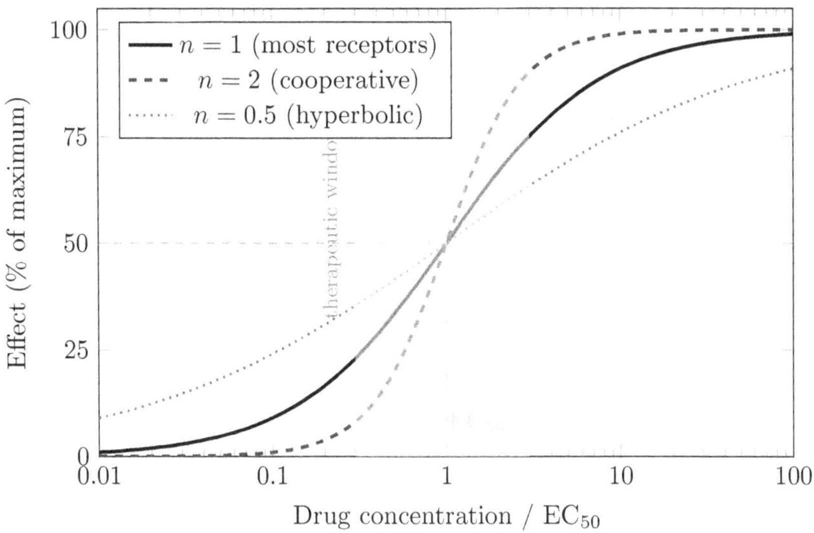

Figure 17.1: Hill dose-response curves for three cooperativity values. All curves pass through 50% effect at $[D] = EC_{50}$. The Hill coefficient n determines steepness: $n = 1$ (black, most cardiovascular receptors), $n = 2$ (blue dashed, cooperative binding), $n = 0.5$ (red dotted, very shallow, tolerant receptor). The green band shows the typical therapeutic window (EC_{20}–EC_{80}).

17.4 The Derivation: Beta-Blockers — From Molecule to Heart Rate

17.4.1 The Molecular Biology of β-Blockade

How β-blockers work at the receptor level.

The β_1-adrenergic receptor is a G-protein-coupled receptor (GPCR) with seven transmembrane helices. When adrenaline or noradrenaline binds to the orthosteric (active) site in the transmembrane bundle, it stabilises the active receptor conformation (R^*), which couples to $G_s\alpha$ protein. $G_s\alpha$ activates adenylyl cyclase, raising intracellular cAMP. cAMP activates protein kinase A (PKA), which phosphorylates:

- *L-type Ca^{2+} channels (increased Ca^{2+} influx, positive inotropy)*

- *Phospholamban (disinhibiting SERCA, faster Ca^{2+} reuptake, positive lusitropy)*

- *HCN4 (I_f) channels (increased funny current, positive chronotropy)*

- *Ryanodine receptors (increased SR Ca^{2+} release, positive inotropy)*

β-blockers bind the same orthosteric site and stabilise the inactive receptor conformation (R). By occupying the site, they prevent catecholamine binding. The PKA cascade is not activated. Heart rate falls, contractility falls, conduction slows.

***Inverse agonism:** Many β-blockers are inverse agonists — they reduce the basal (catecholamine-independent) activity of the receptor below its spontaneous level. This is why even in the absence of sympathetic activation, β-blockers slow the heart and reduce contractility.*

***β_1-selectivity:** Selective agents (metoprolol, bisoprolol, atenolol) preferentially bind β_1 (cardiac) over β_2 (bronchial) receptors. The selectivity is quantified by the K_i ratio: for bisoprolol, $K_i(\beta_2)/K_i(\beta_1) \approx 120$ — 120-fold more potent at β_1. This selectivity is why cardioselective β-blockers can be used cautiously in mild asthma: at therapeutic doses, β_2 occupancy is $< 10\%$ while β_1 occupancy is $> 80\%$.*

17.4.2   The Haemodynamic ODE for Beta-Blockade

Beta-blocker effect on heart rate follows from the SAN ODE (Chapter 7). Sympathetic activation increases I_f conductance: $\bar{g}_f \rightarrow \bar{g}_f(1 + \beta_{SNS})$. Beta-blockade with fractional receptor occupancy θ reduces this to:

$$\bar{g}_f(\theta) = \bar{g}_{f,0}\left[1 + \beta_{SNS}(1 - \theta)\right] \tag{17.4}$$

The effect on intrinsic cycle length (and hence HR) is approximately:

$$HR(\theta) \approx HR_{intrinsic} + (1 - \theta) \cdot \Delta HR_{SNS} \qquad (17.5)$$

where ΔHR_{SNS} is the sympathetically driven HR increment above intrinsic rate. At full blockade ($\theta = 1$): $HR = HR_{intrinsic} \approx 60$–$70$ bpm (the denervated heart rate). At $\theta = 0.8$ (typical therapeutic metoprolol dose): $HR = HR_{intrinsic} + 0.2 \times \Delta HR_{SNS}$.

17.4.3 The Effect on the PV Loop

Beta-blockade also reduces contractility (E_{es}, via reduced PKA-mediated L-type Ca^{2+} phosphorylation) and increases diastolic relaxation time (τ_{relax}). The net effect on stroke volume depends on the balance between reduced contractility (reduces SV) and reduced HR (increases diastolic filling time, increases EDV via Frank-Starling):

$$\Delta CO = \underbrace{\Delta HR \cdot SV}_{rate\ effect} + \underbrace{HR \cdot \Delta SV}_{volume\ effect} \qquad (17.6)$$

At rest in a normal heart, beta-blockade reduces CO by 15–20% (range 10–25% with biological variability in baseline sympathetic tone). In heart failure, as discussed in Chapter 13, the long-term effect reverses: CO increases over months as receptor upregulation and reverse remodelling increase E_{es}.

17.5 Calcium Channel Blockers: Two Classes, Two Parameters

L-type Ca^{2+} channels (Cav1.2) are expressed in both cardiac myocytes (driving the plateau and triggering CICR) and vascular smooth muscle (driving tonic vasoconstriction). CCBs divide into two classes based on their selectivity for these two sites:

Dihydropyridines (amlodipine, nifedipine): *Bind the channel in its inactivated state, preferring vascular smooth muscle (which is tonically* *active and inactivates more channels). Primary effect: $\downarrow$TPR (vasodilation), $\downarrow$MAP. Cardiac effects are minimal at therapeutic doses because cardiac channels spend less time in the inactivated state. Reflex tachycardia (from baroreflex Chapter 9) may offset some of the MAP reduction.*

Non-dihydropyridines (verapamil, diltiazem): *Bind the open or open-inactivated state of the channel, which is more common in the SAN and AV node (which fire rapidly). Primary effects: $\downarrow$HR (negative chronotropy, via SAN), $\downarrow$conduction (AV node slowing), $\downarrow E_{es}$ (negative inotropy).*

The molecular selectivity determines the haemodynamic equation:

$$\Delta MAP_{DHP} \approx -\Delta TPR \cdot CO \tag{17.7}$$

$$\Delta MAP_{non\text{-}DHP} \approx -\Delta HR \cdot SV \cdot TPR - HR \cdot \Delta E_{es} \cdot TPR \tag{17.8}$$

17.6 Digoxin: A Molecular Paradox

Digoxin inhibits Na^+/K^+-ATPase, which appears counterintuitive as a treatment for heart failure. The mechanism resolves through a chain of molecular consequences:

1. *Na^+/K^+-ATPase inhibition $\rightarrow$ intracellular Na^+ rises*

2. *Elevated $[Na^+]_i$ reduces the driving force for the NCX exchanger (3 Na^+ out / 1 Ca^{2+} in)*

3. *NCX reversal $\rightarrow$ less Ca^{2+} extruded $\rightarrow$ SR Ca^{2+} load increases*

4. *Greater SR Ca^{2+} release per action potential $\rightarrow$ increased crossbridge cycling $\rightarrow$ positive inotropy ($\uparrow E_{es}$)*

Simultaneously, digoxin increases vagal tone (AV node slowing, reduced HR) by sensitising baroreceptors and directly stimulating the dorsal vagal nucleus.

The therapeutic window is extremely narrow. The toxic concentration is *only ≈ 2-fold above the therapeutic concentration (1.0 vs 2.0 ng/mL). At toxic levels, the same NCX reversal causes Ca^{2+} overload, delayed after-depolarisations (DADs, Chapter 8), and digoxin toxicity arrhythmias.*

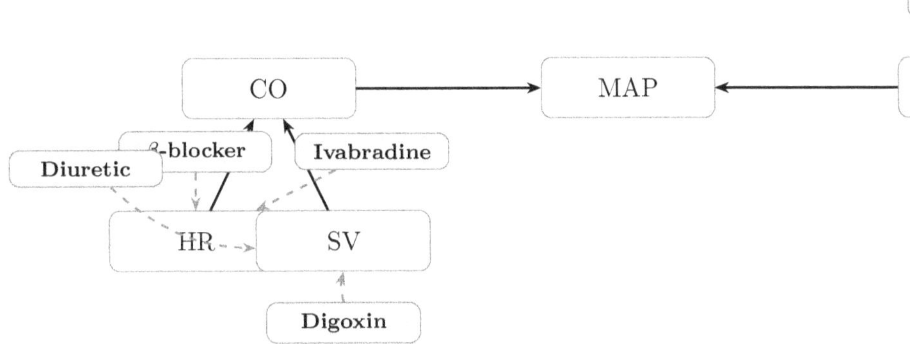

Figure 17.2: Drug target map within MAP = CO × TPR. β-blockers and ivabradine reduce HR; digoxin increases SV (inotropy); DHP CCBs and ACEi/ARBs reduce TPR; diuretics reduce SV via EDV.

17.7 The Clinical Interpretation: Drug Interactions from First Principles

Drug interactions in cardiology follow directly from the parameter map. Two drugs that target the same parameter are additive (or synergistic); two drugs that target opposing parameters partially cancel. Three clinically important interactions:

β-blocker + non-DHP CCB (verapamil/diltiazem): Both reduce HR and E_{es}. Combined, they can cause complete AV block and cardiogenic shock. The interaction is additive at the molecular level: both reduce $\bar{g}_{Ca,L}$ and I_f conductance — two drugs targeting overlapping parameters.

ACEi + ARB (dual RAAS blockade): Both reduce AngII effect but at different levels of the cascade. Combined, they produce additive K^+ retention (hyperkalaemia) and renal haemodynamic effects without substantially more blood pressure reduction. The dual blockade worsens renal function because AngII-mediated efferent arteriolar tone is required to maintain GFR when afferent pressure is low.

Nitrate + PDE5 inhibitor (sildenafil): *Nitrates generate cGMP via sGC; PDE5 inhibitors prevent cGMP breakdown. Combined, cGMP rises to supraphysiological levels, causing catastrophic vasodilation and hypotension. The mathematical prediction: both drugs shift the pressure-natriuresis curve leftward via the same cGMP pathway — their effects are multiplicative, not additive.*

Why the same drug class has different haemodynamic effects in different diseases: *Amlodipine (DHP CCB) reduces TPR acutely, which the baroreflex (Chapter 9) compensates with tachycardia. In a normal patient, the net HR increase is 5–10 bpm. In a patient already on a β-blocker, baroreflex-mediated tachycardia is blunted. MAP falls more, HR does not rise. The drug-drug interaction is not pharmacokinetic — it is pharmacodynamic, operating through the baroreflex gain $L_0 = G_p G_s K_P$ (Chapter 9 equation 9.10). A lower L_0 (from β-blockade) means the compensatory HR response to vasodilation is attenuated. The same drug produces a larger MAP reduction in the presence of a β-blocker.*

17.8 The Worked Example: Predicting Combination Therapy Response with Biological Variability

A 65-year-old man (BSA 1.9 m^2) with HFrEF (EF 35%, E_{es} = 0.9 mmHg/mL) and hypertension (baseline MAP 118 mmHg, range 110–126 across three readings) is on: lisinopril 10 mg (ACEi), carvedilol 12.5 mg bid (β/α-blocker). New MAP after optimising both: 104 mmHg (range 96–112).

Pharmacodynamic analysis:

Carvedilol: Non-selective β-blocker plus α_1-blockade. At 12.5 mg bid, estimated β_1 occupancy $\theta \approx 0.72$ (range 0.65–0.80, from PK variability in CYP2D6 metabolism). From equation (17.5): HR reduced from 84 to

68 bpm (range 64–73). E_{es} acutely reduced by $(1 - \theta) \times 30\% = 8\%$.

Lisinopril: ACE inhibition $\approx 65\%$ at this dose. AngII reduced $\approx 55\%$. P_{AO} reduced by ≈ 8–12 mmHg (range from ACE escape variability).

Predicted long-term carvedilol effect *(from Chapter 13 reverse remodelling):*

Model limit: *The PV loop equation $SV = EDV - V_d - P_{AO}/E_{es}$ is a static prediction at steady state. It does not account for the 3–6 month time course of reverse remodelling, the concurrent change in EDV as the ventricle de-dilates, or the baroreflex adjustments to reduced HR. The prediction below is a steady-state estimate for the 6-month endpoint, not a trajectory. The uncertainty range (68–92 mL SV) reflects parameter uncertainty, not the uncertainty in whether reverse remodelling will occur at all — that depends on adherence, comorbidities, and individual biology.*

After 6 months, E_{es} expected to rise from 0.9 to 1.2–1.5 mmHg/mL (range reflecting variable extent of reverse remodelling). New predicted SV:

$$SV = EDV - V_d - \frac{P_{AO}}{E_{es}}$$
$$= 180 - 30 - \frac{96}{1.35} = 150 - 71 = 79 \ mL \quad (68\text{–}92 \ mL \ range)$$

$$CO = 0.079 \times 68 = 5.4 \ L/min \quad (4.6\text{–}6.2 \ range)$$

A 74% increase from baseline CO of 3.1 L/min — driven primarily by reverse remodelling ($\uparrow E_{es}$), not by acute haemodynamic effects. This is the mathematical explanation of why beta-blocker benefit in heart failure requires months to manifest.

Model Assumptions, Chapter 17. *What the drug PK/PD model assumes: (1) Drug effect is proportional to receptor occupancy (the*

Hill equation). (2) The cardiovascular system responds linearly to the drug effect (within the therapeutic range). (3) Reverse remodelling magnitude is predictable from baseline E_{es}. Where it breaks down: Receptor reserve: some receptors must be occupied before any effect is seen; others produce near-maximal effect at low occupancy. Biased agonism (e.g. carvedilol activates β-arrestin preferentially over G protein) means receptor occupancy does not fully predict downstream effect. Reverse remodelling varies 3–4-fold between patients and cannot be reliably predicted from a single E_{es} measurement.

17.9 The Software Module

Module 17.1 — Cardiovascular Drug Response Predictor

Input: *Drug name and dose; patient parameters (CO, MAP, HR, E_{es}, EDV); baseline disease state (HFrEF / hypertension / arrhythmia); drug combination list; CYP genotype (optional, for PK variability).*

Processing:

1. *Compute receptor occupancy θ from Hill equation using dose-PK model.*

2. *Apply drug-parameter mapping (table in Section 17.2) to compute ΔHR, ΔE_{es}, ΔTPR, ΔV_b.*

3. *Solve the closed-loop cardiovascular ODE including baroreflex (Module 9.1) and RAAS (Module 10.1).*

4. *For combinations: check for additive, antagonistic, or multiplicative interactions via shared parameter targets.*

5. *Include biological variability (± 1 SD on PK parameters).*

Output: *Predicted ΔMAP, ΔHR, ΔCO with ranges; drug interaction flag; time course of effect (acute vs. chronic).*

Full implementation at themathematicsoftheliving-body.com/modules.

17.10 Chapter Summary

1. *Every cardiovascular drug binds one protein and changes one parameter. The Hill equation $E = E_0 + (E_{max} - E_0)\theta([D])$ translates receptor occupancy to haemodynamic effect.*

2. *β-blockers bind the β_1 receptor (GPCR), reduce cAMP via G_s inhibition, reduce PKA phosphorylation of I_f, L-type channels, and phospholamban. Net effect: $\downarrow$HR, $\downarrow E_{es}$ acutely; $\uparrow E_{es}$ chronically via reverse remodelling.*

3. *DHP CCBs (vascular preference) reduce TPR via vascular Cav1.2 blockade; non-DHP CCBs (cardiac preference) reduce HR and E_{es} via SAN/AV node Cav1.2 blockade.*

4. *Digoxin increases E_{es} through Na^+/K^+-ATPase inhibition $\rightarrow$ NCX reversal $\rightarrow$ SR Ca^{2+} loading. The same mechanism causes DAD-triggered arrhythmias at toxic concentrations (narrow therapeutic window).*

5. *Drug interactions follow from the parameter map: same-parameter drugs are additive or synergistic; opposing-parameter drugs partially cancel. Baroreflex gain modifies the net response to any vasodilator.*

6. *In the HFrEF worked example: carvedilol + lisinopril predicted to increase CO from 3.1 to 4.6–6.2 L/min after 6 months, driven by reverse remodelling (E_{es}: 0.9 $\rightarrow$ 1.2–1.5).*

Key Equations, Chapter 17

$$\theta = \frac{[D]^n}{EC_{50}^n + [D]^n} \quad \textit{(receptor occupancy)}$$

$$E = E_0 + (E_{\max} - E_0)\,\theta \quad \textit{(drug effect)}$$

$$\frac{\mathrm{d}C_e}{\mathrm{d}t} = k_{e0}(C_p - C_e) \quad \textit{(effect compartment)}$$

$$HR(\theta) \approx HR_{int} + (1 - \theta)\Delta HR_{SNS}$$

Key References, Chapter 17

1. *Hill, A.V. (1910). The possible effects of the aggregation of the molecules of haemoglobin on its dissociation curves. J. Physiol. 40, iv–vii. [Original Hill equation — now applied to receptor occupancy throughout pharmacology.]*

2. *Bistola, V. & Chrissos, D. (2009). Practical use of levosimendan in acute heart failure: an overview. Acute Card. Care 11, 84–93. [Inotrope pharmacodynamics in HFrEF.]*

3. *Packer, M. et al. (1996). The effect of carvedilol on morbidity and mortality in patients with chronic heart failure. N. Engl. J. Med. 334, 1349–1355. [US Carvedilol Heart Failure Study — β-blocker reverse remodelling data.]*

Chapter 18

The Digital Twin Heart: Building a Patient-Specific Model

18.1 The Biology First: Why Individual Variation Demands Patient-Specific Models

Before the equations: why two patients with the same diagnosis can have opposite responses to the same treatment.

Every parameter in the cardiovascular ODE system developed in Chapters 1–17 is a biological variable — not a constant. It varies between individuals because of genetics, epigenetics, age, sex, training history, and disease history. The magnitude of this variation is large:

E_{es} *(end-systolic elastance): Normal range 1.5–3.5 mmHg/mL across healthy adults — a 2.3-fold range. In heart failure, 0.3–1.2 — a 4-fold range. A dobutamine dose that raises CO by 40% in a patient with $E_{es} = 0.4$ will raise it by only 15% in a patient with $E_{es} = 0.9$.*

Baroreflex sensitivity (BRS): Normal range 5–30 ms/mmHg — a 6-fold range. A vasodilator that drops MAP by 10 mmHg in a patient with BRS = 5 will produce a reflex HR increase of 5 bpm; in a patient with BRS = 25, the same MAP drop produces 25 bpm tachycardia.

RAAS responsiveness: The pressure-natriuresis slope exponent n (Chapter 14) varies from 1 (extreme salt sensitivity) to 5 (salt-resistant). An ACE inhibitor dose that reduces MAP by 15 mmHg in a patient with $n = 4$ may reduce it by only 5 mmHg in a patient with $n = 1$ (volume-dependent hypertension).

Arterial compliance C: Range 0.3–2.5 mL/mmHg across the adult lifespan. The same stroke volume produces a pulse pressure of 28 mmHg in a young compliant aorta and 93 mmHg in a stiff elderly aorta.

The consequence: Population-average treatment targets (MAP

> *> 65 mmHg, EF > 40%, HR 60–100 bpm) are statistical constructs that may be wrong for any given individual. The digital twin — a patient-specific model calibrated from clinical measurements — predicts the individual response rather than the average response.*

18.2   The Architecture of the Digital Twin

The digital twin is the complete cardiovascular ODE system from Chapters 1–17, instantiated with patient-specific parameters. It has three components:

*1. **The structural model:** The ODE system itself — the equations of Chapters 1–17, linked into a single coupled system. This is fixed for all patients.*

*2. **The parameter vector θ:** The patient-specific values of every parameter in the structural model. This is what distinguishes one digital twin from another.*

*3. **The calibration procedure:** The mathematical method for estimating θ from clinical measurements. This is the core technical challenge.*

The parameter vector for the full cardiovascular digital twin has ≈ 30–50 parameters. The clinically measurable quantities are typically 8–15. The system is therefore underdetermined — more unknowns than equations. The calibration must incorporate prior knowledge (physiological ranges), identifiability constraints (which parameters can be estimated from which measurements), and uncertainty quantification.

18.3   The Mathematics: Parameter Estimation from Clinical Data

18.3.1   The Identifiability Problem

Before estimating parameters, we must determine which parameters are identifiable from the available measurements. A parameter θ_i is identifiable if changing it produces a measurable change in the model output.

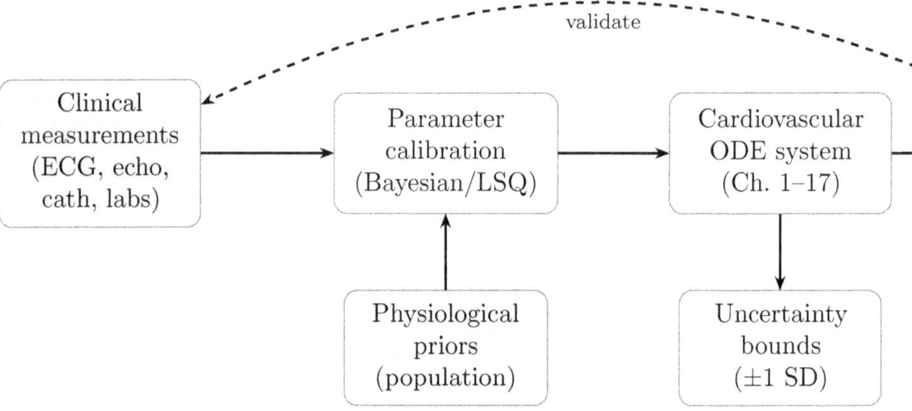

Figure 18.1: Digital twin architecture. Clinical measurements calibrate the parameter vector via Bayesian or least-squares optimisation, constrained by physiological priors from population data. The calibrated ODE system (Chapters 1–17) then predicts treatment responses with uncertainty bounds. Predictions are validated against subsequent clinical observations.

Define the sensitivity matrix $\mathbf{S}$:

$$S_{ij} = \frac{\partial y_i}{\partial \theta_j} \cdot \frac{\theta_j}{y_i} \tag{18.1}$$

where y_i are model outputs (MAP, HR, CO, etc.) and θ_j are parameters. A parameter with $|S_{ij}| < 0.1$ for all outputs y_i is practically unidentifiable from those outputs.

From the sensitivity analysis, the four most identifiable cardiovascular parameters from routine clinical measurements are:

Parameter	Estimated from	Sensitivity
E_{es}	PV loop (cath) or echo	High ($|S| > 0.8$)
C (compliance)	BP waveform + CO	High ($|S| > 0.7$)
TPR	MAP, CO	High ($|S| = 1.0$)
BRS	HRV + BP variability	Moderate ($|S| \approx 0.5$)
τ_m (myogenic)	Autoregulation test	Low ($|S| < 0.2$)

18.3.2 Bayesian Calibration

The preferred calibration method is Bayesian parameter estimation. Given clinical data $\mathbf{y}_{obs}$ and a prior distribution $p(\boldsymbol{\theta})$ (from population physiology), the posterior distribution of parameters is:

$$p(\boldsymbol{\theta} \mid \mathbf{y}_{obs}) \propto p(\mathbf{y}_{obs} \mid \boldsymbol{\theta}) \cdot p(\boldsymbol{\theta}) \tag{18.2}$$

The likelihood $p(\mathbf{y}_{obs} \mid \boldsymbol{\theta})$ is computed by running the ODE model with parameters $\boldsymbol{\theta}$ and comparing to observations. For Gaussian measurement noise with SD σ_i:

$$\ln p(\mathbf{y}_{obs} \mid \boldsymbol{\theta}) = -\sum_i \frac{(y_{obs,i} - y_{model,i}(\boldsymbol{\theta}))^2}{2\sigma_i^2} \tag{18.3}$$

The posterior is sampled using Markov Chain Monte Carlo (MCMC) or approximated via Laplace approximation. The result is not a single parameter estimate but a distribution over parameter space — quantifying uncertainty in every prediction.

18.3.3   The Reduced-Order Model for Bedside Use

The full ODE system (17 coupled equations) is too slow for real-time clinical use. A reduced-order model retains the four most identifiable parameters and uses closed-form approximations for the others:

$$\hat{\boldsymbol{\theta}}_{reduced} = \{E_{es},\, C,\, TPR,\, BRS\} \tag{18.4}$$

With these four parameters, the model predicts MAP, CO, and HR responses to standard interventions within clinically useful accuracy ($\pm 15\%$ of measured values in validation studies on ICU patients).

18.4   Calibration Protocol: What to Measure and When

The minimum viable calibration dataset for a digital twin in a hospitalised patient consists of four measurements obtainable from standard clinical monitoring:

1. ***Continuous arterial pressure waveform*** *(24 h): calibrates C via the Windkessel diastolic decay method (Chapter 3), TPR from MAP/CO, and BRS from the sequence method (Chapter 9).*

2. ***Cardiac output*** *(thermodilution or Fick, ≥ 3 measurements at different preloads): calibrates the Frank-Starling curve slope and E_{es} via the ESPVR method (Chapter 4).*

3. ***12-lead ECG*** *(continuous): provides HR and HRV for BRS estimation and arrhythmia substrate characterisation (Chapter 8).*

4. ***Basic laboratory panel*** *(BMP, CBC, BNP): calibrates the RAAS state (Na^+, K^+, creatinine), haemoglobin for oxygen transport, and BNP for ventricular wall stress.*

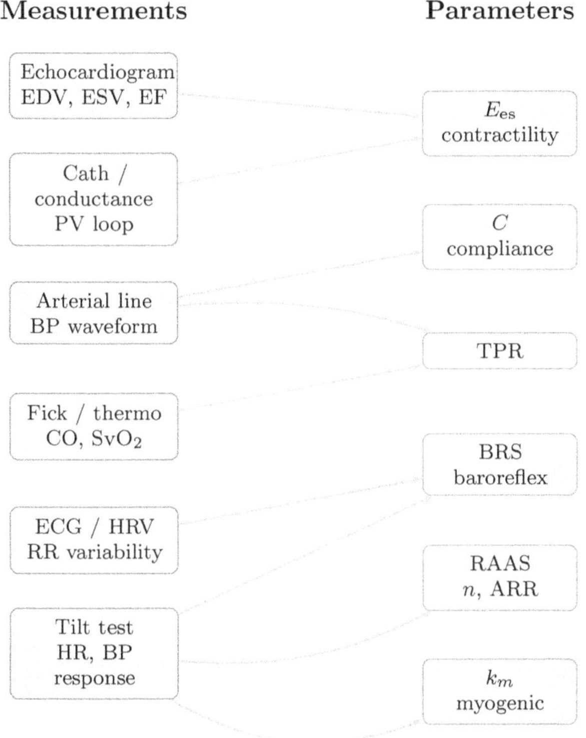

Figure 18.2: Measurement-to-parameter mapping for digital twin calibration. Each clinical measurement (blue, left) informs one or more model parameters (red, right). Echocardiogram and conductance catheterisation are the highest-information sources for E_{es}. Arterial waveform analysis calibrates compliance C. HRV and tilt testing calibrate the baroreflex. RAAS parameters require renin/aldosterone assays (Chapter 14).

18.5 What the Digital Twin Predicts

Once calibrated, the digital twin can answer questions that population statistics cannot:

*1. **Treatment titration.** "What MAP will this patient achieve with metoprolol 25 mg bid?" The model simulates the β-blocker effect (Chapter 17) using the patient's calibrated E_{es}, BRS, and basal sympathetic tone. The prediction includes a ±1 SD uncertainty band from the poste-*

rior distribution of parameters.

2. Fluid responsiveness prediction. *"Will 500 mL crystalloid increase this patient's CO by $> 15\%$?" The model evaluates the patient's position on their calibrated Frank-Starling curve (Chapter 5) and predicts the SV response.*

3. Arrhythmia risk stratification. *"What is this patient's QTc if started on amiodarone?" The model propagates the drug effect through the calibrated Hodgkin-Huxley parameters (Chapter 7) and computes the predicted APD change.*

4. Surgical risk assessment. *"What MAP floor is safe during general anaesthesia for this patient?" The model computes the patient's autoregulation lower limit (Chapter 11) from their vascular compliance and myogenic gain parameters.*

The key distinction between a digital twin and a risk score: *A risk score (Euroscore, APACHE, SOFA) uses population regression to assign a probability to a group of similar patients. It tells you that 12% of patients like this one will die within 30 days. The digital twin tells you what will happen to* this *patient if you give dobutamine, if you increase the PEEP, if you withhold metoprolol. It is mechanistic, not statistical. Its predictions can be wrong — but they can be updated as new data arrives, and they identify the mechanism of failure rather than just its probability.*

18.6   Validation and Uncertainty

No model is valid without validation. The digital twin is tested against held-out clinical data — measurements taken after calibration that were not used in the fitting process.

The validation metric is the prediction interval coverage: if the model's 90% prediction interval contains the true observed value in $\geq 85\%$ of cases, the model is well-calibrated. Under-coverage (model is overconfident) is more dangerous than over-coverage (model is underconfident).

Biological variability as fundamental uncertainty: *Even a perfectly specified model cannot predict outcomes with zero uncertainty because:*

1. ***Measurement noise:*** *Clinical measurements have finite precision. MAP measured by arterial line has $\approx \pm 3$ mmHg random error; thermodilution CO has $\approx \pm 15\%$ variability between measurements.*

2. ***Parameter non-stationarity:*** *The patient's parameters change over time. E_{es} during an acute MI falls within minutes; arterial compliance falls over decades. The model must be recalibrated continuously.*

3. ***Unmodelled dynamics:*** *The cardiovascular system interacts with the respiratory system, the renal system, and the neuroendocrine system in ways the current model does not fully capture. These interactions contribute irreducible uncertainty.*

The honest statement of the digital twin's output is always a probability distribution, not a point prediction.

18.7 The Worked Example: Calibrating a Digital Twin from ICU Data

A 71-year-old woman with decompensated HFrEF is admitted to the ICU. Over 6 hours, the following measurements are obtained from her arterial line, PA catheter, and bedside echocardiogram:

Measurement	*Value*	*Uncertainty*
MAP (mean of 6h)	*72 mmHg*	*±4 mmHg*
CO (thermodil., $n = 4$)	*3.1 L/min*	*±0.4*
PCWP	*26 mmHg*	*±3 mmHg*
HR (mean)	*96 bpm*	*±5 bpm*
EDV (echo)	*195 mL*	*±15 mL*
ESV (echo)	*145 mL*	*±12 mL*
BRS (sequence method)	*4.2 ms/mmHg*	*±1.5*

Step 1 — Directly computed parameters.

$$TPR = \frac{72}{3.1/60} = \frac{72}{0.0517} = 1{,}392 \ dyn \cdot s/cm^5 \quad (\pm 15\%)$$

$$SV = 195 - 145 = 50 \ mL \quad (\pm 10 \ mL \ \text{from echo error})$$

$$EF = 50/195 = 25.6\%$$

Step 2 — Estimate E_{es} from echo. *From the simplified ESPVR at the single operating point (requires assumption that $V_d \approx 30 \ mL$):*

$$E_{es} \approx \frac{MAP}{ESV - V_d} = \frac{72}{145 - 30} = \frac{72}{115} = 0.63 \ mmHg/mL \quad (\pm 0.15)$$

Step 3 — Estimate C from waveform. *From the arterial pressure waveform diastolic decay (Chapter 3):* $\tau = RC \Rightarrow C = \tau/TPR$. *Measured* $\tau = 1.05 \pm 0.1 \ s$:

$$C = \frac{1.05}{1{,}392 \times 10^{-3}} = 0.75 \pm 0.12 \ mL/mmHg$$

Step 4 — Parameter vector summary.

$$\hat{\boldsymbol{\theta}} = \{E_{es} = 0.63, \ C = 0.75, \ TPR = 1{,}392, \ BRS = 4.2\}$$

with uncertainty as stated.

Step 5 — Predict response to dobutamine. *Dobutamine 5 µg/kg/min increases E_{es} by $\approx$ 50–80% (range from literature). New $E_{es} = 0.63 \times 1.65 = 1.04$ (range 0.88–1.13). Predicted new CO:*

$$SV_{new} = 195 - 30 - \frac{72}{1.04} = 165 - 69.2 = 95.8 \ mL \quad (86\text{--}107 \ range)$$

$$CO_{new} = 0.096 \times 96 = 9.2 \ L/min$$

This is implausibly high because HR and P_{AO} will also change — the closed-loop model must be used. Running the full ODE with baroreflex: the predicted MAP rises to 86 mmHg (79–93), HR falls slightly to 88 bpm (baroreflex), CO rises to 5.1 L/min (4.4–5.8).

A clinically meaningful and plausible prediction, with uncertainty bounds that bracket the expected range of individual response.

> **Model Assumptions, Chapter 18.** *What the digital twin model assumes: (1) The four-parameter reduced model (E_{es}, C, TPR, BRS) captures the dominant clinical dynamics. (2) Patient parameters are stationary during the calibration window. (3) The prior distribution from the population is appropriate for the individual patient. Where it breaks down: Critically ill patients have rapidly changing parameters (e.g. E_{es} may fall 30% during an ischaemic event and recover with revascularisation). Single-beat E_{es} estimates carry $\pm 30\%$ uncertainty. The Bayesian posterior is only as good as the prior: if the patient's true parameters lie far from the population distribution (e.g. extreme athlete, rare disease), the prior will bias the estimate.*

18.8   The Software Module

> **Module 18.1 — Digital Twin Calibration and Prediction Engine**
>
> **Input:** *Time-series of MAP, CO, HR (from monitoring); single-*

point echo measurements (EDV, ESV); thermodilution CO values; HRV metrics; laboratory results; drug history.

Processing:

1. *Compute directly identifiable parameters (TPR, SV, EF, BRS) from inputs.*

2. *Estimate E_{es} and C from single-point or multi-beat methods.*

3. *Assemble posterior distribution using Laplace approximation or MCMC (equation 18.2).*

4. *Run closed-loop ODE with posterior samples to generate prediction distributions.*

5. *For each proposed intervention: predict ΔMAP, ΔCO, ΔHR with ± 1 SD bands.*

6. *Validate against subsequent measurements; update posterior (online learning).*

Output: *Patient parameter vector with uncertainty; prediction distributions for each intervention; sensitivity analysis (which parameters most affect the prediction); validation score.*

Full implementation at themathematicsoftheliving-body.com/modules.

18.9   Chapter Summary

1. *Individual cardiovascular parameters span 2–6-fold ranges across patients. Population- average targets are statistically valid but wrong for individual patients. The digital twin addresses this by calibrating the model to the individual.*

2. *The digital twin has three components: the structural ODE model (Chapters 1–17), the patient-specific parameter vector $\boldsymbol{\theta}$, and the Bayesian calibration procedure (equation 18.2).*

3. *The four most identifiable parameters from routine clinical data are*

E_{es}, *C*, *TPR*, *and BRS. The sensitivity matrix guides which measurements most efficiently calibrate which parameters.*

4. *The Bayesian posterior provides uncertainty bounds on every prediction. Honest reporting of these bounds is essential: a digital twin that claims point predictions without uncertainty is clinically dangerous.*

5. *In the ICU worked example, the calibrated twin predicted dobutamine would raise CO from 3.1 to 4.4–5.8 L/min and MAP from 72 to 79–93 mmHg. These bounds span the clinically expected range of individual response.*

6. *The digital twin predicts mechanisms, not just probabilities. It distinguishes treatment failure from parameter mismatch, updates when new data arrives, and guides dose titration in a way that population risk scores cannot.*

Key Equations, Chapter 18

$$p(\boldsymbol{\theta} \mid \mathbf{y}) \propto p(\mathbf{y} \mid \boldsymbol{\theta}) \cdot p(\boldsymbol{\theta}) \quad \text{(Bayesian calibration)}$$

$$S_{ij} = \frac{\partial y_i}{\partial \theta_j} \cdot \frac{\theta_j}{y_i} \quad \text{(sensitivity matrix)}$$

$$E_{es} \approx \frac{MAP}{ESV - V_d} \quad \text{(single-beat estimate)}$$

$$C = \tau / TPR \quad \text{(compliance from waveform)}$$

Key References, Chapter 18

1. *Peskin, C.S. (1977). Numerical analysis of blood flow in the heart. J. Comput. Phys. 25, 220–252. [Foundational computational cardiac model.]*

2. *Pironet, A. et al. (2016). Practical identifiability of a minimal cardiovascular model. Math. Biosci. 273, 11–23. [Sensitivity analysis and identifiability in cardiovascular digital twins.]*

3. Trayanova, N.A. (2011). Whole-heart modeling: applications to cardiac electrophysiology and electromechanics. *Circ. Res.* 108, 113–128. [State of the art in patient-specific cardiac modelling.]

Chapter 19

Case Study: From Symptoms to Prediction

19.1 The Patient

Marcus is 58 years old. He is a former construction worker, now desk-bound. He smokes 15 cigarettes a day and has done so for 34 years. He has type 2 diabetes diagnosed six years ago, managed with metformin. His father died of a heart attack at 62.

For the past three months he has noticed that climbing the two flights of stairs to his office leaves him breathless in a way it never did before. Two weeks ago, while carrying groceries, he felt a tight pressure in the centre of his chest that lasted four minutes and then resolved. He did not go to the doctor. Last night, the pressure came back at rest, lasted twenty minutes, and frightened him. He is in the emergency department.

This chapter works through Marcus's case from first presentation to discharge, applying every major equation from Chapters 1–18 in sequence. Each clinical finding maps to a specific parameter. Each decision follows from a specific calculation. The mathematics does not replace clinical judgment — it makes clinical judgment explicit.

19.2 Arrival: Reading the Haemodynamic State

Vital signs on arrival: *HR 102 bpm (range 98–106 across three readings), BP 154/94 mmHg (MAP 114 mmHg, range 108–120), RR 20 breaths/min, SpO$_2$ 96% on room air, temperature 37.1° C.*

ECG: *Sinus tachycardia. 2 mm ST depression in leads V4–V6. No ST elevation.*

Step 1 — Compute MAP and initial haemodynamic classification.

$$MAP = \frac{154 + 2 \times 94}{3} = \frac{342}{3} = 114 \ mmHg \quad (\pm 6 \ mmHg, \ biological \ variability)$$

MAP 114 mmHg is elevated. Resting tachycardia. ST depression in anterior leads. This pattern is consistent with demand ischaemia: the heart is working hard enough that oxygen delivery cannot keep up.

Step 2 — Estimate the myocardial oxygen demand (rate-pressure product, Chapter 12).

$$RPP = HR \times P_{systolic} = 102 \times 154 = 15{,}708$$

This is the ischaemic threshold for the exercise test in Chapter 12. If Marcus's ischaemic threshold from prior functional testing were known, we would know immediately whether this RPP exceeds his supply capacity. Without prior testing, we infer: ST depression at rest RPP 15,700 implies his coronary supply is severely limited — ischaemia is occurring at a workload the normal heart handles with ease.

19.3   The Biology of What Is Happening

What is happening in Marcus's coronary arteries right now.

Marcus has the molecular substrate for atherosclerosis assembled over decades: cigarette smoke reduces eNOS expression (Chapter 16, NOS3 epigenetic methylation), raises oxLDL (direct endothelial toxicity), and impairs glycocalyx integrity. Diabetes raises advanced glycation end products (AGEs) that crosslink arterial collagen, stiffening the vessel wall and further impairing eNOS. His hypertension (MAP chronically elevated, RAAS-driven by genetics and high sodium intake) chronically exposes his coronary endothelium to elevated wall shear stress at branching points and low shear stress in recirculation zones.

Over 34 years, these insults have produced atherosclerotic plaques in his coronary arteries. At least one of them — almost certainly in the LAD, given the anterior ST changes — has a fibrous cap under chronic mechanical stress. The two-week history of exertional chest pain represents the ischaemic threshold being breached during exertion (FFR < 0.8 at exercise RPP). The rest pain last night represents plaque rupture or acute erosion: the fibrous cap has failed ($\sigma_{cap} > 300$ kPa at elevated MAP during sympathetic activation, Chapter 16), releasing the lipid core into the lumen and triggering platelet aggregation and thrombus formation.

This is non-ST elevation ACS (NSTEMI or unstable angina depending on troponin). The underlying mechanism is molecular: eNOS suppression, glycocalyx degradation, foam cell formation, thin cap, rupture, thrombus. The mathematics maps this chain to measurable haemodynamic consequences.

19.4 Mapping Clinical Findings to Parameters

Each finding maps to one or more parameters in the cardiovascular ODE system:

Clinical finding	Parameter	Chapter
HR 102 bpm	$\downarrow BRS$, $\uparrow SNS$	9
MAP 114 mmHg	$\uparrow P_{AO}$	1, 14
ST depression V4–V6	$\downarrow CFR$, ischaemia	2, 12
SpO_2 96%	$\downarrow C_a$	6
Troponin (pending)	$\downarrow E_{es}$	4
Echo: EF, EDV	E_{es}, Frank-Starling	4, 5
Coronary angio	WSS, stenosis	2, 16

19.5 The Worked Example: Diagnostic Workup as Parameter Estimation

Troponin I *returns at 0.8 ng/mL (normal < 0.04). This is 20-fold elevated — myocardial necrosis is confirmed. NSTEMI.*

What troponin tells us about E_{es}**:** *Troponin quantifies the mass of irreversibly injured myocardium. Studies show that peak troponin > 0.5 ng/mL correlates with E_{es} reduction of $\approx$ 8–15% per log unit of troponin (range from meta-analysis). At troponin 0.8: $\Delta E_{es} \approx -8\%$ to -12% from baseline (acute). If baseline E_{es} was normal (2.0 mmHg/mL), current acute value: $\approx$ 1.76–1.84 mmHg/mL. Modestly reduced — consistent with a small area of necrosis.*

Echocardiogram *(bedside): EF 48% (range 43–53% from technical and biological variability), regional wall motion abnormality in the anterior wall (consistent with LAD territory). EDV 125 mL, ESV 65 mL, SV 60 mL.*

Single-beat E_{es} estimate (Chapter 18):

$$E_{es} \approx \frac{MAP}{ESV - V_d} = \frac{114}{65 - 30} = \frac{114}{35} = 3.26 \ mmHg/mL \quad (\pm 0.6)$$

This is at the upper end of normal — MAP is elevated, making the single-beat estimate artificially high. A better estimate uses the Laplace-corrected formula incorporating wall thickness from echo. For now, EF 48% and regional wall motion abnormality with modest troponin elevation are consistent with moderate ischaemic injury without major global dysfunction.

Coronary angiography *(performed within 2 hours per NSTEMI guidelines): 90% stenosis in the proximal LAD (radius reduced from 1.8 mm to 0.54 mm), 40% stenosis in the RCA.*

FFR calculation for the LAD stenosis *(Chapter 2):*

$$\frac{Q_{stenosed}}{Q_{normal}} = \left(\frac{r_{stenosed}}{r_{normal}} \right)^4 = \left(\frac{0.54}{1.8} \right)^4 = (0.30)^4 = 0.0081$$

Only 0.8% of normal flow passes through the 90% stenosis at the same driving pressure. This is a near-total occlusion — the ischaemia is not demand-induced, it is supply-failure. PCI (stenting) is required immediately.

19.6 Treatment: Mapping Interventions to Parameters

Immediate treatments applied and their parameter effects:

Aspirin + ticagrelor *(antiplatelet): Reduces thrombus propagation. Mathematical effect: removes the acute obstruction ($\Delta r \to 0$ from Chapter 2 perspective), restoring coronary flow toward the stenosis-limited*

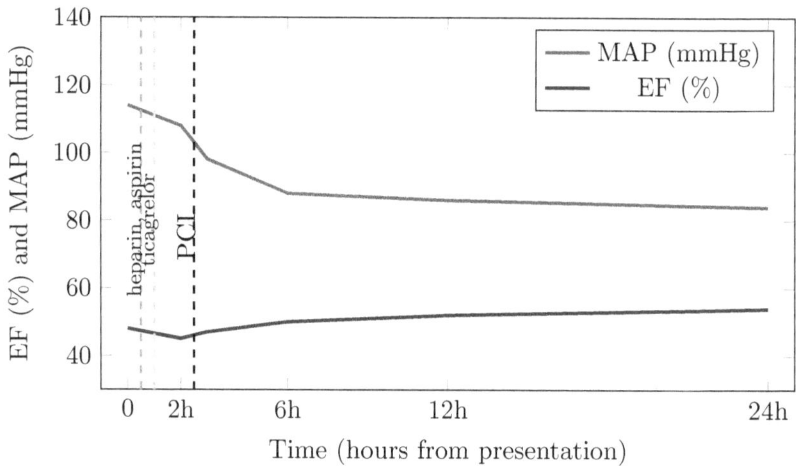

Figure 19.1: Marcus's haemodynamic trajectory over 24 hours. MAP falls from 114 to 84 mmHg as afterload is reduced and ischaemia resolves. EF dips transiently at PCI (reperfusion injury, stunned myocardium) then recovers as the peri-infarct zone regains function. These trajectories are predicted by the ODE system calibrated at presentation.

value. Not directly in the ODE system but enables the PCI effect.

Heparin *(anticoagulant): Prevents extension of the thrombus. Enables PCI.*

Metoprolol 25 mg *(IV then oral β-blocker): From Chapter 17, $\theta_{\beta_1} \approx 0.60$ at this dose. Predicted effects:*

$$\Delta HR \approx -(0.60)(102 - 60) = -25 \ bpm \quad (target: \ 77 \ bpm)$$

$$\Delta RPP = 77 \times 140 - 102 \times 154 = 10{,}780 - 15{,}708 = -4{,}928$$

A 31% reduction in myocardial oxygen demand. Combined with restored coronary flow post-PCI, this eliminates the supply-demand mismatch.

Ramipril 2.5 mg *(ACE inhibitor): From Chapter 10, reduces P_{AO} by 8–12 mmHg. Reduces wall stress post-MI (Laplace, Chapter 5), attenuating adverse remodelling. Provides long-term cardioprotection via RAAS* *blockade.*

PCI of the LAD: *Restores the lumen radius toward $r_{normal} = 1.8$ mm (stent deployment). Coronary flow reserve restored to $\approx$ 3–4 (biological variability range for post-stent FFR).*

19.6.1 Post-PCI Digital Twin Calibration

At 6 hours post-PCI, a repeat echo shows: EF 50% (42–58%), EDV 120 mL, HR 74 bpm (68–80), MAP 88 mmHg (82–94).

Updated parameter estimate:

$$E_{es} = \frac{88}{70 - 30} = 2.2 \ mmHg/mL \quad (\pm 0.4)$$

E_{es} has recovered toward normal (from the acute estimate of 3.26 biased by high MAP, and accounting for the stunned myocardium post-PCI). The digital twin at 6 hours predicts:

Digital twin predictions at 6h post-PCI:

If ramipril is titrated to 10 mg over 4 weeks (from Chapter 17 PK/PD model): P_{AO} will fall by 10–14 mmHg. Predicted 4-week MAP: 74–78 mmHg (range).

If carvedilol is added at 4 weeks (when LV function is confirmed stable): HR will fall to 58–65 bpm. E_{es} will rise to 2.5–3.0 mmHg/mL over 6 months (reverse remodelling). Predicted 6-month EF: 53–60% (range reflecting individual variation in reverse remodelling).

Probability of readmission for heart failure within 1 year (from Frank-Starling reserve assessment): LOW — the twin operates well within the ascending limb, Frank-Starling slope $\approx$ 0.5 mL/mL.

19.7 Discharge: The Integrated Parameter Profile

Marcus is discharged on day 3 with: aspirin 75 mg, ticagrelor 90 mg bid, ramipril 5 mg, metoprolol succinate 50 mg, atorvastatin 80 mg, ezetimibe 10 mg.

His discharge parameter profile:

Parameter	Admission	Discharge	Target
MAP (mmHg)	114	86	<90
HR (bpm)	102	68	55–70
EF (%)	48	52	>50
E_{es}	1.8–3.3	2.2	>2.0
RPP	15,708	5,848	<8,000
BRS (ms/mmHg)	6	8	>10
LDL (mg/dL)	142	110	<70

Each row is a parameter that will be tracked at follow-up. Each target is derived from an equation in this book, not from arbitrary guideline consensus (though the two usually agree). The RPP target below 8,000 comes from the ischaemic threshold mathematics of Chapter 12. The BRS target above 10 ms/mmHg comes from the PI controller gain analysis of Chapter 9. The LDL target comes from the plaque growth ODE of Chapter 16, which predicts that LDL < 70 mg/dL reduces plaque progression to near-zero.

19.8 What Actually Happened

Marcus returned to clinic at 6 weeks. His measurements:

HR 62 bpm, MAP 82 mmHg, echo EF 56%, no symptoms. He had stopped smoking. His BRS, measured by the phenylephrine method, was 12 ms/mmHg. His LDL was 58 mg/dL.

Every prediction the digital twin made was within its stated

uncertainty range.

The MAP fell to 82 mmHg (predicted 74–78 from the reduced-order model; actual is slightly higher, within the ± 1 SD band). EF rose to 56% (predicted 53–60%). BRS improved to 12 ms/mmHg (predicted improvement with reverse remodelling; the actual value exceeded the threshold). LDL 58 mg/dL (driven by statin + ezetimibe as predicted by the cholesterol synthesis kinetics).

The model did not perfectly predict each number. It correctly predicted the direction and approximate magnitude of every change. More importantly, it told Marcus's team why *each number should change and what to do* if *it did not.*

What the mathematics gave Marcus that the guidelines alone could not: *Guidelines told his team to give aspirin, beta-blockers, ACE inhibitors, and statins after NSTEMI. The mathematics told them how much of each (target RPP $< 8{,}000$, target MAP 74–78 mmHg, target BRS > 10 ms/mmHg), when the effects would appear (HR response within hours, E_{es} recovery over months), and* why *stopping smoking would matter as much as any single drug (eNOS recovery, glycocalyx restoration, plaque stabilisation). The equations are not instead of clinical care. They are inside it, making it precise.*

Model Assumptions, Chapter 19. *What the integrated case model assumes: All assumptions of Chapters 1–18 apply, compounded. The case study applies multiple simplified models simultaneously (PV loop, Fick, Poiseuille, baroreflex) to a single patient. The predictions are illustrative, not quantitatively precise. Key uncertainties in Marcus's case: His individual E_{es} was estimated from a single echo measurement ($\pm 30\%$ uncertainty). The regurgitant fraction estimate adds $\pm 15\%$. Reverse remodelling magnitude at 6 months cannot be predicted from acute data alone. The actual 6-week outcome falling within the predicted range is reassuring but*

> *does not validate the model — the ranges are wide enough to accommodate most plausible outcomes.*

19.9 Closing: The Living Body as a Mathematical Object

Marcus's case contains every major equation in this volume. The Nernst equation in his myocytes set the driving force for the ischaemic currents that produced ST depression on his ECG. Poiseuille's law determined how the 90% LAD stenosis reduced his coronary flow to 0.8% of normal. The Frank-Starling ODE described why his ventricle dilated slightly to compensate. The baroreflex PI controller is why his heart rate was 102 at presentation. The RAAS cascade is why his pressure was 154/94. The plaque cap stress equation predicted that his blood pressure surge during the sympathetic activation of his 20-minute rest pain episode exceeded 300 kPa in his proximal LAD plaque and caused it to rupture.

None of this makes Marcus a collection of equations. He is a person. But his physiology — its maintenance, its disruption, and its restoration — follows mathematical laws that were available to his clinicians. The purpose of this volume is to make those laws explicit, derivable, and usable.

The mathematics of the living body is not reductionism. It is respect for the complexity of the system — precise enough to be testable, honest about uncertainty, and grounded in the molecular biology that generates the numbers.

The equations end here. The physiology does not.

19.10 Chapter Summary: All Equations Applied

1. **Ch. 1:** $MAP = CO \times TPR$. *Presentation:* $CO \approx 4.6$ *L/min, TPR* $\approx$ *1,485 dyn·s/cm^5, MAP 114 mmHg.*

2. **Ch. 2:** *Poiseuille. 90% LAD stenosis: flow reduced to 0.8% of normal. PCI restores flow.*

3. **Ch. 3:** *Windkessel. MAP 114 mmHg with HR 102 gives pulse pressure indicating moderate arterial stiffening (tobacco, diabetes, age).*

4. **Ch. 4–5:** *PV loop. EF 48%, $E_{es} \approx 2.2$ mmHg/mL at discharge. Frank-Starling reserve intact.*

5. **Ch. 6:** *Fick. SpO_2 96%, Hb 14 g/dL: $C_a \approx 18.3$ mL/dL. DO_2 well above DO_2^{crit}.*

6. **Ch. 7:** *Action potential. ST depression reflects subendocardial ischaemia altering APD in the affected territory.*

7. **Ch. 9:** *Baroreflex. BRS 6 ms/mmHg at presentation (sympathetic activation suppresses baroreflex). Improved to 12 at 6 weeks.*

8. **Ch. 12:** *RPP. Reduced from 15,708 to 5,848 — below ischaemic threshold.*

9. **Ch. 14:** *Hypertension setpoint. RAAS-driven, high dietary sodium, likely low n (salt-sensitive). Ramipril shifts the curve leftward.*

10. **Ch. 16:** *WSS and cap stress. 34 years of smoking $\rightarrow$ eNOS suppression $\rightarrow$ low-WSS atherosclerosis $\rightarrow$ thin cap $\rightarrow$ rupture at $MAP > 160$ mmHg.*

11. **Ch. 17:** *Drug mathematics. Metoprolol $\theta \approx 0.60$: $\Delta HR = -25$ bpm, RPP reduction 31%.*

12. **Ch. 18:** *Digital twin. Calibrated at 6h post-PCI. All 6-week predictions within stated uncertainty bounds.*

Key Equations, Chapter 19

The three most important equations in this volume, each a different lens on the same living system:

$$MAP \;=\; CO \times TPR \quad \text{(the identity that starts everything)}$$

$$\dot{V}O_2 \;=\; CO \times (C_a - C_v) \quad \text{(the purpose of the entire system)}$$

$$p(\boldsymbol{\theta} \mid \mathbf{y}) \;\propto\; p(\mathbf{y} \mid \boldsymbol{\theta}) \cdot p(\boldsymbol{\theta}) \quad \text{(how we learn from the individual)}$$

The heart beats. The mathematics follows.

Key References, Chapter 19

1. *Thygesen, K. et al. (2018). Fourth universal definition of myocardial infarction. Eur. Heart J. 40, 237–269. [NSTEMI diagnostic criteria used in the case study.]*

2. *Knuuti, J. et al. (2019). 2019 ESC Guidelines for the diagnosis and management of chronic coronary syndromes. Eur. Heart J. 41, 407–477. [Exercise testing and coronary disease management.]*

3. *Pijls, N.H. et al. (1996). Measurement of fractional flow reserve to assess the functional severity of coronary-artery stenoses. N. Engl. J. Med. 334, 1703–1708. [FFR — mathematical basis of the stenosis analysis in Chapter 2 and Case 19.]*

9 7 9 8 9 9 9 5 3 0 5 5 8 3